Communications in Computer and Information Science 3120

Rationale

The CCIS series is devoted to the publication of proceedings of computer science conferences. Its aim is to efficiently disseminate original research results in informatics in printed and electronic form. While the focus is on publication of peer-reviewed full papers presenting mature work, inclusion of reviewed short papers reporting on work in progress is welcome, too. Besides globally relevant meetings with internationally representative program committees guaranteeing a strict peer-reviewing and paper selection process, conferences run by societies or of high regional or national relevance are also considered for publication.

Topics

The topical scope of CCIS spans the entire spectrum of informatics ranging from foundational topics in the theory of computing to information and communications science and technology and a broad variety of interdisciplinary application fields.

Information for Volume Editors and Authors

Publication in CCIS is free of charge. No royalties are paid, however, we offer registered conference participants temporary free access to the online version of the conference proceedings on SpringerLink (http://link.springer.com) by means of an http referrer from the conference website and/or a number of complimentary printed copies, as specified in the official acceptance email of the event.

CCIS proceedings can be published in time for distribution at conferences or as post-proceedings, and delivered in the form of printed books and/or electronically as USBs and/or e-content licenses for accessing proceedings at SpringerLink. Furthermore, CCIS proceedings are included in the CCIS electronic book series hosted in the SpringerLink digital library at http://link.springer.com/bookseries/7899. Conferences publishing in CCIS are allowed to use our online conference service (Meteor) for managing the whole proceedings lifecycle (from submission and reviewing to preparing for publication) free of charge.

Publication process

The language of publication is exclusively English. Authors publishing in CCIS have to sign the Springer CCIS copyright transfer form, however, they are free to use their material published in CCIS for substantially changed, more elaborate subsequent publications elsewhere. For the preparation of the camera-ready papers/files, authors have to strictly adhere to the Springer CCIS Authors' Instructions and are strongly encouraged to use the CCIS LaTeX style files or templates.

Abstracting/Indexing

CCIS is abstracted/indexed in DBLP, Google Scholar, EI-Compendex, Mathematical Reviews, SCImago, Scopus. CCIS volumes are also submitted for the inclusion in ISI Proceedings.

How to start

To start the evaluation of your proposal for inclusion in the CCIS series, please send an e-mail to ccis@springer.com

Haoyu Chen · Zhao Ren · Jiadong Wang ·
Emily Provost · Ehsan Adeli · Björn W. Schuller ·
Tanja Schultz · Guoying Zhao
Editors

Human Behavior Analysis and Interaction for Emotional Intelligence

The First Workshop on EI-MiGA@IJCAI 2026, IJCAI 2026
Bremen, Germany, August 15–21, 2026
Proceedings

Springer

Editors

Haoyu Chen
University of Oulu
Oulu, Finland

Zhao Ren
University of Bremen
Bremen, Germany

Jiadong Wang
Technical University of Munich
Munich, Germany

Emily Provost
University of Michigan
Ann Arbor, MI, USA

Ehsan Adeli
Stanford University
Stanford, CA, USA

Björn W. Schuller
Technical University of Munich
Munich, Germany

Tanja Schultz
University of Bremen
Bremen, Germany

Guoying Zhao
CMVS, University of Oulu
Oulu, Finland

ISSN 1865-0929 ISSN 1865-0937 (electronic)
Communications in Computer and Information Science
ISBN 978-981-92-4528-4 ISBN 978-981-92-4529-1 (eBook)
https://doi.org/10.1007/978-981-92-4529-1

This Springer imprint is published by the registered company Springer Nature Singapore Pte Ltd.
The registered company address is: 152 Beach Road, #21-01/04 Gateway East, Singapore 189721, Singapore

Preface

It is our great pleasure to welcome you to the proceedings of The First Joint Workshop on Human Behavior Analysis and Interaction for Emotional Intelligence, with the 4th MiGA Challenge (EI-MiGA 2026), held on August 17, 2026, in Bremen, Germany, in conjunction with the 35th International Joint Conference on Artificial Intelligence (IJCAI 2026).

Artificial Intelligence is undergoing a profound paradigm shift in how it processes human affect. While traditional affective computing has heavily focused on sensing and categorizing discrete emotional states, next-generation Emotion AI is moving toward holistic understanding, reasoning, and context-aware empathetic interaction. Beyond simply detecting human expressions, modern systems are increasingly required to analyze the underlying intentions and causes of affect, and generate meaningful explanations and nuance-controlled responses. The EI-MiGA workshop was established to provide a unified platform to explore this full spectrum of emotional intelligence in AI.

In tandem with the main workshop track, the 4th Micro-Gesture Analysis for Hidden Emotion Understanding (MiGA) Challenge continued its push toward benchmark-driven evaluation of spontaneous human behaviors. Focusing on micro-gestures which are highly informative yet notoriously difficult to capture automatically, the challenge provided a rigorous proving ground for advanced computer vision and machine learning models tasked with evaluating emotional stress and real-world behavioral states.

Each submission to the workshop underwent a rigorous single-blind peer-review process, with every paper evaluated by at least three members of our international Program Committee. Out of 20 submissions received, the committee selected 13 papers for inclusion in these proceedings, published in the Springer Communications in Computer and Information Science (CCIS) series. The accepted papers represent high-quality original research, insightful viewpoints, and innovative solutions falling within the workshop's scope and the MiGA challenge.

In addition to these peer-reviewed contributions, this volume features a comprehensive position paper authored by the Organizing Committee. This paper serves to summarize the key methodologies and breakthroughs presented across this year's submissions, while offering a strategic roadmap and future insights for the field of emotional intelligence.

We would like to express our deepest gratitude to everyone who contributed to the success of EI-MiGA 2026. First and foremost, we thank the authors for sharing their pioneering work and the challenge participants for pushing the technical boundaries of behavior analysis. We are immensely grateful to the members of our program committee for their meticulous, constructive, and timely reviews. We also extend our thanks to the IJCAI 2026 workshop chairs for their continuous guidance and structural support. Finally, we thank Microsoft CMT for providing the peer-review management platform.
It is our hope that these proceedings serve as a valuable reference, inspiring further

innovation, cross-disciplinary collaboration, and the development of effective, socially responsible, and empathetic AI interfaces.

June 2026

Haoyu Chen
Zhao Ren
Jiadong Wang
Emily Provost
Ehsan Adeli
Björn W. Schuller
Tanja Schultz
Guoying Zhao

Organization

Program Committee Chairs

Haoyu Chen	University of Oulu, Finland
Zhao Ren	University of Bremen, Germany
Jiadong Wang	Technical University of Munich University Hospital, Germany & Munich Center for Machine Learning, Germany
Emily Mower Provost	University of Michigan, USA
Ehsan Adeli	Stanford University, USA
Björn W. Schuller	Imperial College London, UK & Technical University of Munich University Hospital, Germany & Munich Center for Machine Learning, Germany
Tanja Schultz	University of Bremen, Germany
Guoying Zhao	University of Oulu, Finland

Program Committee

Haolin Zuo	Inner Mongolia University, China
Yang Tan	Technical University of Munich University Hospital, Germany
Xianghu Yue	Tianjin University, China
Xin Jing	Technical University of Munich University Hospital, Germany
Xiaoliang Wu	University of Southampton, UK
Yilin Pan	Dalian Maritime University, China
Wei Xue	Xi'an Jiaotong University, China
Xinyuan Qian	University of Science and Technology Beijing, China
Yong Ma	University of Bergen, Norway
Hanliang Xie	Peking University, China
Kun Li	United Arab Emirates University, United Arab Emirates
Chengyan Wang	University of Oulu, Finland
Yueyi Yang	University of Oulu, Finland
Dashanka De Silva	University of Bremen, Germany
Sheng Li	Institute of Science Tokyo, Japan

Yi Chang	Imperial College London, UK
Xiaoyu Yuan	University of Oulu, Finland
Beining Xu	Shenzhen MSU-BIT University, China
Xiaolei Zhong	Kunming University of Science and Technology, China
Huayu Chen	Technical University of Munich, Germany
Jiachen Luo	Queen Mary University of London, UK
Saurav Pahuja	University of Bremen, Germany
Xi Chen	Chinese University of Hong Kong (Shenzhen), China
Zeyu Chen	Technical University of Munich University Hospital, Germany
Changzeng Fu	Northeastern University, China
Zipeng Zhang	Technical University of Munich University Hospital, Germany
Rathi Adarshi Rammohan	University of Bremen, Germany
Xuesong Yang	NVIDIA Corporation, USA
Haixu Liu	University of Sydney, Australia
Yan Jiang	University of Oulu, Finland
Zhaodi Xu	Nanjing University of Posts and Telecommunications, China

Contents

Human Behavior Analysis and Interaction for Emotional Intelligence: Advances, Challenges, and the Road Ahead — EI-MiGA 2026

Haoyu Chen[1(✉)], Zhao Ren[2], Jiadong Wang[3,7], Emily Mower Provost[4], Ehsan Adeli[5], Björn W. Schuller[3,6,7], Tanja Schultz[2], and Guoying Zhao[1]

[1] CMVS, University of Oulu, Finland, Finland
{chen.haoyu,guoying.zhao}@oulu.fi
[2] Cognitive Systems Lab, University of Bremen, Germany, Germany
{zren,tanja.schultz}@uni-bremen.de
[3] CHI, Technical University of Munich University Hospital, Germany, Germany
jiadong.wang@tum.de
[4] University of Michigan, USA, USA
emilykmp@umich.edu
[5] Stanford University, USA, USA
eadeli@stanford.edu
[6] GLAM, Imperial College London, United Kingdom, United Kingdom
bjoern.schuller@imperial.ac.uk
[7] Munich Center for Machine Learning, Germany, Germany

Abstract. Emotional intelligence, the capacity of machines to perceive, interpret, and respond to human affective states, demands a holistic understanding of behavior that spans subtle body movements, spoken language, and social interaction. This paper presents the proceedings and position of *The First Joint Workshop on Human Behavior Analysis and Interaction for Emotional Intelligence* at IJCAI 2026, bringing together with *the 4th MiGA Challenge*. The **competitive MiGA challenge** attracted 66 registered participants from worldwide institutions across three tracks. Beyond the competition session, the contributed workshop papers in the other two sessions further highlight recent advances in emotional intelligence. The **Explainability in Emotional Intelligence** session features papers on interpretability-driven speech emotion recognition and hate speech detection, as well as trustworthiness in speech language models. The **Affective Interaction and Social Computing** session focuses on socially grounded and safety-critical topics, including richness-aware emotional dialogue generation and clinical safety auditing of implicit sycophancy in mental health dialogue systems. Together, these contributions position the field toward trustworthy and socially aware emotional intelligence, while exposing persistent open challenges in evaluation rigor, domain adaptation, and clinical deployment.

H. Chen, Z. Ren and J. Wang: These authors contributed equally to this work.

H. Chen et al. (Eds.): IJCAI 2026, CCIS 3120, pp. 1–14, 2027.
https://doi.org/10.1007/978-981-92-4529-1_1

Keywords: Affective Computing · Emotional Intelligence · Human Behavior Analysis · Micro-Gesture Recognition · Social Computing

1 Introduction

Developing emotionally intelligent machines requires more than standard pattern recognition; it demands a genuine capacity to interpret and understand human emotions. While existing systems focus on detecting discrete or dimensional emotions, future AI will evolve to reason about affective experiences and generate nuanced, empathetic, and interactive responses [1–3]. AI with *Emotional Intelligence* demands a unified account of how affective states are encoded across human behaviors [4–6]. Human behavior is inherently multimodal and temporally layered: the subtle body movements betray suppressed feelings, voice prosody and lexical choices shape conversational affect, physiological signals capture internal affective shifts [7,8], and the social dynamics of interaction amplify or mask individual emotional expression. No single modality or task is sufficient; progress requires coordinated advances across perception, reasoning, and interaction.

The **First EI-MiGA Workshop (Human Behavior Analysis and Interaction for Emotional Intelligence)** is held in IJCAI 2026, together with the **4th MiGA Challenge (Micro-Gesture Analysis for Hidden Emotion Understanding)**, to advance AI technologies in emotional intelligence. The workshop explores the full spectrum of Emotion AI within affective computing, starting with multimodal affect sensing from speech, vision, and physiological biosignals. The goal is to advance affect-aware systems that not only perceive affective states, but also understand and respond to them in an empathetic manner. The workshop also places particular emphasis on model trustworthiness [9], bridging technical innovation with ethical, fair, responsible, and robust human-computer interaction (HCI) [10–12]. Ultimately, the workshop is expected to foster the development of next-generation affect-aware interfaces that are both effective and socially responsible.

The MiGA Challenge series has been established since IJCAI 2023 as MiGA (Micro-Gesture Analysis for Hidden Emotion Understanding) [13–16]. Over three successive editions, the series attracted over 200 cumulative registered participants, progressively raising the bar on micro-gesture classification, online recognition, and behavior-based emotion reasoning. The 2025 winner achieved 69.2 % accuracy on behavior-based emotion prediction, a task that requires inferring hidden emotional polarity purely from visual behavioral cues, without speech or identity information. The MiGA challenge also inspired the community, and more related workshops and competitions have come out [17,18].

The **2026 EI-MiGA Workshop and Challenge**[1] deliberately broadens the scope beyond gesture recognition toward a wider conception of emotional intelligence in analyzing multi-modal human behavior, varying from video, speech, to text modalities. Co-organized by the researchers from the University

[1] https://ei-miga.github.io.

of Oulu, the University of Bremen, Stanford University, the Technical University of Munich, and the University of Michigan, the workshop is structured around three thematic sessions that together span the behavioral, linguistic, and social dimensions of affective AI. **Session 1** (Micro-Gesture Challenge and Recognition) continues the competitive challenge tradition, pushing the state of the art on gesture perception and emotion inference. **Session 2** (Explainability in Emotional Intelligence) addresses AI explainability issues in spoken and textual affective interaction. **Session 3** (Affective Interaction and Social Computing) examines socially grounded and safety-related topics in deployed dialog-based affective systems.

Workshop Itinerary. The EI-MiGA Workshop was jointly organized with the 4th MiGA challenge, which was hosted on the Kaggle platform with three independent leaderboards[2,3,4]. The overall timeline of the workshop and challenge was as follows:

- **Apr. 10, 2026.** The call for the challenge was announced, and registration opened.
- **Apr. 13, 2026.** The training data, development toolkit, and sample codes were released.
- **May 6, 2026.** The testing data were released.
- **May 9, 2026.** The final test submission deadline was reached, and registration closed.
- **May 10, 2026.** The challenge results were released.
- **May 26, 2026.** The paper submission deadline.
- **Jun. 01, 2026.** Notification to authors.
- **Jun. 12, 2026.** Camera-ready deadline.
- **Aug. 17, 2026.** The EI-MiGA IJCAI 2026 Workshop was held in Bremen, Germany.

Below, we will introduce the advances, recurring challenges, and forward-looking directions that emerge across all three sessions.

2 Session 1: The MiGA Challenge

2.1 Competition Tracks and Datasets

The MiGA 2026 challenge consists of three competitive tracks addressing complementary aspects of micro-gesture-based affective understanding.

Track 1 – Micro-gesture Classification requires fine-grained recognition of 4,563 pre-segmented test clips into 32 predefined micro-gesture categories on the iMiGUE dataset [14], with pretrained models on the iMiGUE3K dataset [19], evaluated by top-1 accuracy. The **iMiGUE** [14] dataset consists of 18,499 samples across 32 micro-gesture categories plus a non-gesture class.

[2] https://www.kaggle.com/competitions/the-4th-ei-mi-ga-ijcai-challenge-track-1.
[3] https://www.kaggle.com/competitions/the-4th-ei-mi-ga-ijcai-challenge-track-2.
[4] https://www.kaggle.com/competitions/the-4th-ei-mi-ga-ijcai-challenge-track-3.

Track 2 – Online Micro-gesture Recognition requires joint temporal localization and fine-grained classification of micro-gesture instances in long untrimmed video sequences on the SMG dataset [13], evaluated by F1 score combining precision and recall over detected segments, with baseline toolkits provided [20,21]. The **SMG** [13,15] dataset captures spontaneous micro-gestures under psychological stress, comprising 3,692 samples across 17 fine-grained categories from 40 subjects, with four available modalities: RGB video, skeleton, depth, and optical flow.
Track 3 – Behavior-based Hidden Emotion Prediction requires inferring hidden emotional polarity (match win/loss) from muted, identity-anonymized post-match interview videos of professional tennis players on the iMiGUE dataset [14], using only visual behavioral cues, from 359 interview sequences of professional tennis players captured under competitive pressure.

Those datasets are all collected in an unconstrained manner with in-the-wild performed body movements during daily communications, posing substantial challenges for affective behavior modeling, including pronounced class imbalance, high semantic similarity across categories, and extremely subtle motion patterns.

A total of **66** entrants registered across all three tracks: **25** for Track 1, **18** for Track 2, and **23** for Track 3. A total of **34** teams were verified for the final evaluation stage, with **553** submissions made across all tracks during the challenge period.

2.2 Challenge Results and Methods

The validity of each top-ranked solution was verified by requiring teams to submit source code and test-set predictions.

Track 1: Multimodal MG Classification. Table 1 summarizes the top-3 methods on Track 1. The 2026 results reveal a major methodological shift toward self-supervised pretraining and lightweight class-wise expert fusion.

Table 1. Top-3 methods for Track 1 (MG Classification, iMiGUE).

Team	Rank	Modality	Key Technique	Acc.
XInsight Lab [22]	1	Skel.+RGB+Depth	MG-FM-RGB (SSL) + MM-Gesture ens.	**74.42 %**
YUV [23]	2	Multimodal (12 exp.)	CLOG-Fusion: class-wise log-opinion fusion	**74.24 %**
Xidian [24]	3	Skeleton + RGB	SkateFormer + IB-Memory + JBMA + ViViT	**71.37 %**

The **first-place method** (XInsight Lab) [22] proposes a self-supervised ensemble that integrates MG-FM-RGB—a SMILE-based masked video modeling framework pretrained on 120K unlabeled in-domain iMiGUE clips without action-level annotations—with the supervised multimodal MM-Gesture pipeline from prior editions [25]. The self-supervised RGB stream alone achieves

69.22 % top-1 accuracy, demonstrating substantial transferability of representations learned from unlabeled in-domain data. The final ensemble reaches **74.42 %**, surpassing prior state-of-the-art by 1.21 percentage points and establishing self-supervised in-domain pretraining as a new standard paradigm for micro-gesture recognition.

The **second-place method** (YUV, Harbin Institute of Technology Shenzhen) [23] presents CLOG-Fusion, a 416-parameter class-wise log-opinion fusion head over a SwinT-enhanced pool of 12 frozen expert models—spanning RGB, Taylor-series video, optical flow, motion-region crops, MA-52-pretrained RGB-Trans variants, and PoseC3D skeleton branches. Rather than training a new backbone, CLOG-Fusion learns a class-specific simplex weighting over stored expert logits, reflecting that expert reliability is inherently category-dependent in fine-grained gesture recognition. The system achieves **74.24 %**, ranking second with substantially fewer trainable parameters.

The **third-place method** (Xidian University) [24] targets discriminative skeleton representation learning. Information Bottleneck Memory (IB-Memory) maintains class-specific memory banks to suppress performer-specific jitter and task-irrelevant variations by distilling robust motion semantics. Joint-Bone Mutual Alignment (JBMA) enforces bidirectional KL consistency between joint and bone prediction distributions to enhance semantic complementarity before feature-level fusion. A ViViT RGB branch provides complementary appearance cues, integrated via temperature-calibrated decision-level fusion, reaching **71.37 %** top-1 accuracy.

Beyond the top-3, a notable fourth-place entry (HFUT and University of Auckland) [26] proposed a MultiModal Cross-Subject Pseudo-Labeling framework that addresses the cross-subject domain gap, achieving **68.13 %** accuracy, which is a competitive performance compared to the last year.

Track 2: Online MG Recognition. Table 2 summarizes the top-3 methods on Track 2.

Table 2. Top-3 methods for Track 2 (Online MG Recognition, SMG).

Team	Rank	Modality	Key Technique	F1
XInsight Lab [27]	1	RGB (PEFT)	ST-Decoupled Adapter + VideoMAE + DyFADet	**0.43808**
AIM [28]	2	RGB + Skeleton	DyFADet+ with Gated Residual Fusion	**0.40880**
XD-L [29]	3	Skeleton only	Skeleton Temporal Det. + Add-on MLP Refiner	**0.35302**

The **first-place method** (XInsight Lab) [27] introduces a Spatial-Temporal Decoupled Adapter (STDA) for parameter-efficient fine-tuning of a frozen VideoMAE backbone. Unlike existing adapter-based PEFT methods that process spatial and temporal information within a shared pathway, STDA decomposes video adaptation into parallel spatial and temporal branches via lightweight depthwise convolutions, enabling each to specialize before merging complementary representations. Adaptive Soft Balanced Augmentation (ASBA) further dynamically determines per-class augmentation intensity based on effective sample counts and learning difficulty without manual thresholds, mitigating the severe long-tail class imbalance in the SMG dataset. The system achieves an F1 of **0.4381**, substantially improving upon the 2025 best of 0.380.

The **second-place method** (AIM) [28] extends DyFADet into DyFADet+, a dual-stream RGB-Skeleton framework. Both modalities are projected into shared multi-scale temporal embeddings; a gated residual fusion module then adaptively injects skeleton motion cues into RGB features at each feature pyramid level, conditioned on both modalities to learn contextually where kinematic information is reliable. This design preserves discriminative RGB appearance while selectively reinforcing motion-sensitive regions, achieving **F1 of 0.4088**.

The **third-place method** (XD-L) [29] addresses the precision-recall trade-off in skeleton-only online recognition via a lightweight two-stage candidate refinement strategy. A first-stage skeleton temporal detector produces reliable high-precision base predictions from 525-dimensional frame-level skeleton features. An Add-on MLP refiner subsequently selects from a low-threshold extra candidate pool—deciding whether each candidate complements the base prediction set based on candidate confidence, duration, and overlap features—achieving a public leaderboard score of **0.3530**.

Track 3: Behavior-based Hidden Emotion Prediction. Table 3 summarizes the top-3 results on Track 3.

Table 3. Top-3 methods for Track 3 (Hidden Emotion Prediction, iMiGUE).

Team	Rank	Acc.
Miss Qiu/ XInsight Lab [30]	1	**76.92 %**
AIM	2	**75.96 %**
Jason Karpeles	3	**64.42 %**

The **first-place method** (Miss Qiu / XInsight Lab) [30] presents ReFELS (Rethinking Feature Engineering and Learning Strategies), a compact multimodal temporal framework that reformulates behavior-based hidden emotion prediction as *offset-aware multimodal temporal modeling* rather than generic video classification. The feature representation is carefully engineered to maximize behavioral signal-to-noise ratio: 2D/3D body and face landmarks (OpenPose, MediaPipe), 52-dimensional facial expression blendshapes, DINOv2/DINOv3 frame-level visual features, X-CLIP video embeddings, Gemini-based semantic behavioral priors, optical flow descriptors, and monocular depth features (DA3). Static pose coordinates serve as "Base features" encoding absolute positions, while frame-wise numerical differences construct "Offset features" capturing first-order micro-motion kinematics—shifting model attention from static identity and body-shape biases toward localized relative micro-movement velocities. Architecturally, a cross-attention mechanism uses Base features as Query and Offset features as Key/Value; adaptive Multiple Instance Learning (MIL)-based pooling extracts instantaneous emotional evidence from long sequences. The method achieves **76.92 % accuracy**, a 7.7-point improvement over the 2025 winner [16].

Notably, ReFELS also exposes the "representation collapse" phenomenon of general vision foundation models applied to micro-dynamic tasks, and demonstrates that performance gains in VLM-ensembled pipelines may reflect shortcut learning and rote pseudo-label memorization—"pseudo-generalization"—rather than genuine behavioral understanding. These findings provide critical guidance for rigorous evaluation design in future editions.

The second-place team (AIM) achieved an accuracy of **75.96 %**, and the third-place team (Jason Karpeles) achieved **64.42 %**. As neither team submitted a workshop paper, their methods are not discussed further in this report.

3 Session 2: Explainability in Emotional Intelligence

The **explainability** of artificial intelligence is essential to fostering user trust and comprehension in affective computing and mental healthcare systems [31,32]. This section provides an overview of the three accepted papers that investigate this critical axis within intelligent systems.

Temporal Concept Alignment in Speech Emotion Recognition. The authors in [33] address a critical gap in explainable AI for Speech Emotion Recognition (SER): while visualization methods can highlight salient speech segments driving model decisions, they rarely assess whether these segments align with human expert knowledge of how emotion is expressed in speech. The authors introduce TempA, a computational framework that aligns model-inferred contributing segments (extracted from Concept Relevance Propagation (CRP) saliency maps) with human expert knowledge, including loudness, shrillness, frequency variation, amplitude variation, pitch level, and breathiness. TempA achieves this alignment through a temporal overlap metric inspired by Intersection-over-Union, measuring whether the segments the model attends to coincide with those that vocal emotion theory identifies as emotionally informative.

With TempA, four findings are discovered: 1) Among all tested emotions, frequency and amplitude variation are the acoustic features that the model relies on most heavily, and they also align most closely with expert theoretical frameworks. 2) Shrillness and loudness are extremely important in human acoustic theory. However, they are scarcely referenced in the model's decision-making process. 3) Across different emotions, the patterns of overlap between the model's referenced regions and expert theoretical regions vary dramatically. 4) Not all class-wise CRP heatmaps are worth using for alignment analysis. For the "disgust" emotion, removing the highlighted heatmap regions does not affect the model's judgment.

Hate Speech Classification with Sentence-Level Analysis. In [34], the authors present a multimodal web application that detects hate speech in spoken and written content while prioritizing explainability. The system employs hand-crafted acoustic features and linguistic features for unimodal and multimodal classification. A key innovation locates on two-stage classification: 1) sentence-level classification with neighboring sentences as context (up to 40 % of preceding sentences), 2) full-text classification using the same feature processing and two sentence-level statistics features. With this diagram, the model shows explainability on the sentence-level contribution and additionally provides feature visualization within each sentence.

With the proposed method, sentence-level modeling with 40 % preceding context improves full-text classification F1 from 0.73 to 0.92. Edge case analysis reveals two scenarios where the model performs poorly: 1) when the feature differences between adjacent sentences are substantial; 2) when the predicted labels of adjacent sentences suddenly flip.

Time-Step Alignment on Speech Large Language Models. When processing speech and text using speech Large Language Models (LLMs), a temporal mismatch problem often arises among the acoustic encoder, the speech-token interface, and the LLM. Such a mismatch makes it difficult to map audio time steps directly to model outputs, such as transcripts and answers. The work in [35] presents an overview of advanced studies in LLM-based time-step alignment (TSA-LLM) across audio time, linguistic units, and LLM steps. Six families of relevant approaches are discussed: 1) monotonic ASR alignment scaffolds, 2) forced alignment and long-form timestamping, 3) text-aligned speech-token interfaces, 4) LLM-based forced alignment and word timestamping, 5) timestamp-grounded reasoning and captioning, and 6) retrieval or contextual-biasing methods for rare entities. Based on these six methodological families, the authors argue that a speech LLM should output an "alignment contract", a public interface for each grounded output unit. Furthermore, this survey highlights that multiple metrics should be employed to evaluate a TSA-LLM method, including boundary and span accuracy, grounded faithfulness, calibration, drift and long-form stability, latency and revision, and error stratification. Finally, the survey discusses a hybrid alignment framework that combines ASR-derived alignment with LLM-derived timestamp generation. Ultimately, TSA-LLM approaches hold the potential to increase the trustworthiness of speech LLMs, which is essential for emotional intelligence, particularly in building trust among clinicians and patients in mental healthcare applications.

4 Session 3: Affective Interaction and Social Computing

Effective emotional dialogue generation relies on advanced multimodal fusion techniques that can adapt to changing communication cues, while simultaneously requiring rigorous evaluation frameworks to ensure these systems remain clinically safe and supportive.

Multimodal Fusion in Emotional Dialogue Generation. With the development of computational resource and available data, multimodal fusion shows increasing importance, as features in different modalities are complementary. In [36], the authors tackle a fundamental limitation in multimodal Emotional Dialogue Generation (EDG): the widespread assumption that all modalities contribute uniformly to affective understanding. Their proposed framework, Richness-aware Modality-Role Assignment (RaMRA), reformulates multimodal fusion as an instance-level modality-role assignment problem. Rather than treating text as a fixed backbone with audio and vision as auxiliary refinements, or conversely, predesignating a single non-text modality as primary, RaMRA dynamically estimates the semantic-sentiment informativeness of text, audio, and vision for each dialogue context and assigns dominant/auxiliary roles accordingly. The dominant modality serves as the sole query source for cross-modal attention, while auxiliary streams contribute evidence only through value vectors modulated by their informativeness weights. This asymmetric, role-conditioned routing distinguishes RaMRA from prior dynamic fusion approaches that reweight contributions through scalar gates yet still permit every modality to participate symmetrically in attention. Experiments on IEMOCAP [37] and MELD [38] demonstrate consistent improvements over strong baselines across fluency, semantic alignment, and emotion consistency. The framework proves beneficial both in standard decoder-only settings and when integrated with omnimodal LLMs such as Qwen2.5-Omni [39].

Auditing Implicit Sycophancy in Mental-Health Dialogue. The work in [40] investigates implicit sycophancy in mental-health dialogue, where model responses may sound superficially empathetic but ultimately reinforce harmful user beliefs like catastrophizing and hopelessness. To address this, the authors propose Dynamic Emotional Signature Graphs (DESG), a structured audit framework that integrates LLM-based state extraction, a DESG trajectory graph, and clinical direction evaluation. First, LLMs extract semantic, affective, and cognitive-distortion states from the dialogue. These states are then used to construct the DESG trajectory graph and evaluate whether a response steers the user in a clinically productive, neutral, or harmful direction. The paper introduces two benchmarks for evaluation: 1) a clean, 1,500-window single-response benchmark matched with explicit metadata, lexical features, and others, and 2) a synthetic $3\times1{,}000$-dialogue-window cross-domain diagnostic benchmark. Evaluated on these datasets, their proposed DESG-StateRisk achieves state-of-the-art performance in identifying harmful risks, outperforming non-DESG baselines.

5 Conclusion and Perspectives

The contributed papers advance emotional intelligence research in line with the motivation and goals of our workshop. Nevertheless, significant challenges remain before affective computing can truly achieve intelligence. Taken together, the three sessions of EI-MiGA 2026 paint a coherent picture of where emotional intelligence research stands in 2026, and where it must go. To this end, we outline several promising directions for future research.

5.1 Summary

Session 1 reveals two parallel paradigm shifts in micro-gesture perception. In Track 1, self-supervised in-domain pretraining emerges as a decisive advantage: learning from 120K unlabeled gesture clips without action-level labels surpasses all prior supervised pipelines, signaling that the field is entering a data-scaling regime for gesture representation [22]. Simultaneously, class-wise expert fusion (CLOG-Fusion) demonstrates that 416 trainable parameters over frozen specialist logits can nearly match the self-supervised ensemble, validating lightweight calibration as a practical alternative for resource-constrained deployments [23]. In Track 2, parameter-efficient fine-tuning of frozen video foundation models (ST-Decoupled Adapter) delivers a 15-point F1 gain over 2025, establishing PEFT as the dominant paradigm for online recognition in data-scarce settings [27]. In Track 3, physically grounded feature decomposition (ReFELS) achieves a 7.7-point accuracy improvement while exposing a critical methodological hazard: VLM-augmented pipelines frequently exhibit "pseudo-generalization", where gains reflect shortcut learning rather than genuine behavioral understanding [30]. This finding has broad implications for evaluation design across all affective computing benchmarks.

Session 2 brings together three distinct but complementary approaches to advancing explainability and alignment in spoken and affective language systems. First, the study in [33] presents TempA, a framework that audits whether SER models align with human vocal emotion theory, revealing that models heavily rely on frequency and amplitude variations while critically neglecting human indicators like shrillness and loudness. Moving from acoustic theory to structural explainability, the work in [34] implements a two-stage sentence-level multimodal framework for hate speech detection that incorporates local contextual windows, significantly boosting full-text classification F1-scores to 0.92 while mapping clear sentence-level feature contributions. Finally, the survey in [35] discusses the emerging domain of time-step alignment in speech LLMs (TSA-LLM) to address the temporal mismatches between acoustic encoders and text outputs, proposing an "alignment contract" interface and a multi-metric evaluation standard to ensure grounding. Together, these works emphasize that true trustworthiness in affective and mental health systems requires a tighter coupling between model-inferred outputs, temporal grounding, and human expert knowledge.

Section 3 highlights a dual focus on advancing multimodal emotional dialogue systems and establishing rigorous auditing mechanisms to ensure their clinical safety. First, the study in [36] addresses a key limitation in standard multimodal fusion, the assumption of uniform modality contributions, by proposing RaMRA. This framework dynamically assigns asymmetric dominant or auxiliary roles to text, audio, and vision based on instance-level informativeness, significantly improving fluency and emotional consistency on standard benchmarks and omni-modal LLMs. Second, another study [40] address the safety risks of these systems by investigating implicit sycophancy in mental-health contexts. DESG is introduced as a structured auditing framework that extracts dialogue states to trace emotional trajectories, outperforming traditional baselines in identifying clinically harmful responses across two new benchmarks. Catching the implicit sycophancy requires explicit modeling of clinical-state trajectories, a result with immediate relevance for the deployment of AI counseling and crisis support tools.

5.2 The Road Ahead

EI-MiGA 2026 exposes three cross-cutting challenges that will shape the next generation of research.

First, *evaluation integrity*: the pseudo-generalization phenomenon in Track 3, the alignment gap in TempA, and the shortcut-sensitivity in DESG all point to a systemic weakness in how affective AI systems are validated. Rigorous evaluation protocols with explicit controls for leakage, shortcut learning, and distribution shift are urgently needed.

Second, *explainability and intervention*: contributions across all three sessions demonstrate that performance and explainability are not intrinsically in tension, but that achieving both requires purposeful design. Furthermore, the loop from explanation to intervention is broken, and as long as it remains so, explainability will be confined to post-hoc auditing rather than becoming an integral part of model development and deployment. Therefore, we propose Explainability-Driven Training Objectives. We envision a paradigm shift where interpretability metrics are no longer computed solely for human consumption but are directly incorporated into the training loop as optimization targets. This moves beyond the common practice of "training first, explaining later" toward a new regime of explainability-aware learning.

Third, *the behavioral-language-social integration gap*: while Session 1 advances gesture perception, Sessions 2 and 3 reveal that spoken language, dialogue, and social context carry complementary affective signals. Future editions of EI-MiGA will explicitly target their integration, moving toward systems capable of holistic, multi-channel emotional intelligence in naturalistic interaction. For instance, future work can leverage state-tracking graphs (similar to DESG) to explicitly audit why a model prioritized a specific modality.

Acknowledgment. This work is partially funded by the DFG, German Research Foundation, through the project "Silent Paralinguistics" with grant number 40301193; by the DFG's Reinhart Koselleck Project 442218748 (AUDI0NOMOUS); by the EU H2020 project No. 101135556 (INDUX-R). This work was supported by the Research Council of Finland project (grant 348765), flagship Profi 7 (grant 352788), EmotionAI (grants 336116, 345122, 359854), Research Fellow (371019), EU HORIZON-MSCA-SE ACMod (101130271), University of Oulu, and Infotech Oulu.

References

1. Lian, Z., ET AL.: AffectGPT: a new dataset, model, and benchmark for emotion understanding with multimodal large language models. In: Proceedings of ICML (2025)
2. Lian, Z., et al.: EmoPrefer: can large language models understand human emotion preferences?. In: Proceedings of ICLR (2026)
3. Lian, Z., et al.: Open-vocabulary multimodal emotion recognition: dataset, metric, and benchmark. In: Proceedings of ICML (2025)
4. Schuller, B. W., Batliner, A.: Computational paralinguistics: emotion, affect and personality in speech and language processing, John Wiley & Sons (2013)
5. Järvelä, S., Zhao, G., Nguyen, A., Chen, H.: Hybrid intelligence: human–AI coevolution and learning. British J. Educat. Technol. **56** (2025)
6. Yang, Y., et al.: Saynext-Bench: why do LLMs struggle with next-utterance prediction? (2026). arXiv:2602.00327 arXiv preprint
7. Ren, Z., et al.: An Introduction to Silent Paralinguistics (2025). ArXiv preprint: 2508.18127
8. Pistrosch, S., et al.: Affect decoding in phonated and silent speech production from surface EMG (2026). ArXiv preprint: 2603.11715
9. Ren, Z., Schuller, B.W., Eskofier, B.M., Nguyen, T.N., Nejdl, W.: Guest editorial trustworthy and collaborative AI for personalised healthcare through edge-of-things. IEEE J. Biomed. Health Inform., 5213–5215 (2023)
10. Chang, Y., et al.: STAA-Net: a sparse and transferable adversarial attack for speech emotion recognition. IEEE Trans. Affect. Comput. **16** (2024)
11. Z. Ren, K. Qian, T. Schultz, B. W. Schuller, An overview of the ICASSP special session on AI security and privacy in speech and audio processing, in: Proceedings of ACM Multimedia Asia workshop, 2023
12. Chang, Y., Laridi, S., Ren, Z., Palmer, G., Schuller, B. W., Fisichella, M.: A two-stage defence for robust federated speech emotion recognition. In: Proceedings INTERSPEECH (2026)
13. Chen, H., Shi, H., Liu, X., Li, X., Zhao, G.: SMG: a micro-gesture dataset towards spontaneous body gestures for emotional stress state analysis. Int. J. Comput. Vision **131**, 1346–1366 (2023)
14. Liu, X., Shi, H., Chen, H., Yu, Z., Li, X. Zhao, G.: iMiGUE: an identity-free video dataset for micro-gesture understanding and emotion analysis. In: Proceedings of CVPR, pp. 10631–10642 (2021)
15. Chen, H., Schuller, B. W., Adeli, E., Zhao, G.: The 2nd challenge on micro-gesture analysis for hidden emotion understanding (MiGA) 2024: Dataset and results. In: Proceedings of IJCAI Workshops (2024)
16. Chen, H., Schuller, B. W., Adeli, E., Zhao, G.: The 3rd challenge on human behavior analysis for emotion understanding (MiGA) 2025: From recognition to emotion understanding. In: Proceedings of IJCAI Workshops (2025)
17. Guo, D., et al.: Mac 2024: micro-action analysis grand challenge. In: Proceedings of the 32nd ACM International Conference on Multimedia, pp. 11304–11305 (2024)
18. Li, K., et al.: Mac 2025: the 2nd micro-action analysis grand challenge. In: Proceedings of the 33rd ACM International Conference on Multimedia, pp. 14216–14221 (2025)
19. Wang, C., Chen, H., Wei, H., Yang, Y., Chen, Y., Zhao, G.: iMiGUE-3K: a large-scale benchmark for micro-gesture analysis with self-supervised learning. arXiv preprint arXiv:2605.17179 (2026)

20. Chen, H., Liu, X., Shi, J., Zhao, G.: Temporal hierarchical dictionary guided decoding for online gesture segmentation and recognition. IEEE Trans. Image Process. **29**, 9689–9702 (2020)
21. Chen, H., Liu, X., Zhao, G.: Temporal hierarchical dictionary with HMM for fast gesture recognition. In: 2018 24th International Conference on Pattern Recognition (ICPR), IEEE Computer Society (2018)
22. Liu, T., et al.: Self-supervised learning matters: a simple ensemble solution for micro-gesture recognition. In: Proceedings of the IJCAI 2026 EI-MiGA Workshop (2026)
23. Wang, T., Ma, H., Xia, W., Zhang, J., Huang, J., Xu, Y.: SwinT-enhanced class-wise fusion for micro-gesture classification. In: Proceedings of the IJCAI 2026 EI-MiGA Workshop (2026)
24. Wu, K., Li, Y.: Skeleton-based micro-gesture recognition via information bottleneck memory and joint-bone mutual alignment. In: Proceedings of the IJCAI 2026 EI-MiGA Workshop (2026)
25. Gu, J., Wang, F., Li, K., Wei, Y., Wu, Z., Guo, D.: MM-Gesture: towards precise micro-gesture recognition through multimodal fusion. In: Proceedings of the IJCAI 2025 MiGA Workshop (2025)
26. Zhang, H., Zhang, H., Liu, P., Zhang, Y., Xue, W., Hao, Y.: A multi-modal framework with cross-subject pseudo-labeling and semantic alignment for micro-gesture recognition. In: Proceedings of the IJCAI 2026 EI-MiGA Workshop (2026)
27. Shen, X., Li, K., Wang, F., Qian, W., Jiang, J., Guo, D.: Spatial-temporal decoupled adapter for micro-gesture online recognition. In: Proceedings of the IJCAI 2026 EI-MiGA Workshop (2026)
28. Liu, J., He, X., Liu, P., Shi, J., Zang, H., Hao, Y.: Motion reinforces appearance: RGB-skeleton gated residual fusion for micro-gesture online recognition. In: Proceedings of the IJCAI 2026 EI-MiGA Workshop (2026)
29. Liu, Z., Xu, J., Li, Y.: A lightweight skeleton candidate refinement framework for track 2 online micro-gesture recognition. In: Proceedings of the IJCAI 2026 EI-MiGA Workshop (2026)
30. Guo, X., et al.: Rethinking the role of feature engineering and learning strategies in few-shot hidden emotion recognition. In: Proceedings of the IJCAI 2026 EI-MiGA Workshop (2026)
31. Rammohan, R. A., Ren, Z., Swiderska, A., Küster, D., Schultz, T.: Unveiling deep speech embeddings: acoustic insights into hatespeech detection. In: ITG, pp. 16–20 (2025)
32. Sun, Y., et al.: Label correction enhanced dual-stream multiple instance learning for weakly-supervised depression detection in speech. In: Proceedings INTERSPEECH (2026)
33. Nasr, S., Ren, Z., Murindanyi, S., Johnson, D. S.: TempA: temporal alignment of human expert-referenced with model-inferred acoustic concepts for speech emotion recognition. In: Proceedings of the IJCAI 2026 EI-MiGA Workshop (2026)
34. Rammohan, R. A., Behrendt, J., Immoor, J., Santos, P. F., Küster, D., Schultz, T.: HateScan: a multimodal web application for hate speech classification with sentence-level analysis. In: Proceedings of the IJCAI 2026 EI-MiGA Workshop (2026)
35. Li, S., Li, J., Shinozaki, T.: TSA-LLM: a survey of time-step alignment on speech large language models. In: Proceedings of the IJCAI 2026 EI-MiGA Workshop (2026)

36. Yu, M., Jiang, C., Hu, Z. Zhang, W., Wang, L.: Richness-aware modality-role assignment for multimodal emotional dialogue generation. In: Proceedings of the IJCAI 2026 EI-MiGA Workshop (2026)
37. Busso, C., et al.: IEMOCAP: interactive emotional dyadic motion capture database. Lang. Resources Evaluat. **42** (2008)
38. Poria, S., Hazarika, D., Majumder, N., Naik, G., Cambria, E., Mihalcea, R.: MELD: a multimodal multi-party dataset for emotion recognition in conversations. In: Proceedings of ACL, pp. 527–536 (2019)
39. Xu, J., et al.: Qwen2.5-omni technical report (2025). ArXiv preprint: 2503.20215
40. Han, T., Xu, B., Zhang, H., Lu, Y.: Auditing implicit sycophancy in mental-health dialogue: structured clinical-state diagnostics and clean matched benchmarks. In: Proceedings of the IJCAI 2026 EI-MiGA Workshop (2026)

Session 1: Micro-Gesture Challenge and Recognition

Self-supervised Learning Matters: A Simple Ensemble Solution for Micro-gesture Recognition

Tingyi Liu[1], Kun Li[2(✉)], Fei Wang[1,3,4], Junjie Chen[1,3], Zhiliang Wu[5], Jihao Gu[6], Haixu Liu[7,8], and Dan Guo[1,3]

[1] Hefei University of Technology, Hefei, China
[2] United Arab Emirates University, Al Ain, United Arab Emirates
kunli.hfut@gmail.com
[3] Institute of Artificial Intelligence, Hefei Comprehensive National Science Center, Hefei, China
[4] Anhui Evolution Technology Co., Ltd., Hefei, China
[5] Nanyang Technological University, Singapore, Singapore
[6] University College London, London, UK
[7] The University of Sydney, Sydney, NSW, Australia
[8] Beijing QBoson Quantum Technology Co., Ltd., Beijing, China

Abstract. In this paper, we present XInsight Lab's solution to the micro-gesture classification track of the 4th MiGA Challenge at IJCAI 2026, in which our solution ranked first and achieved a new state-of-the-art result. We propose a multimodal ensemble framework that integrates a self-supervised RGB-based model with supervised multi-stream models from previous solutions. The self-supervised RGB model is pretrained on 120K unlabeled clips via masked video modeling and then fine-tuned on iMiGUE. This simple yet effective RGB baseline achieves 69.224% top-1 accuracy on the iMiGUE test set, demonstrating the benefit of learning transferable representations from unlabeled in-domain videos. By incorporating this model as a complementary branch, the final ensemble reaches 74.419% top-1 accuracy, surpassing the previous state of the art by 1.206% points. Experimental results on iMiGUE, including ablation studies on the ensemble strategy, validate the effectiveness of self-supervised RGB representation learning for micro-gesture recognition.

Keywords: Micro-gesture recognition · Self-supervised pretraining · Affective computing

1 Introduction

Micro-gestures (MGs) [3,27,34] are subtle, spontaneous body movements that may reflect hidden cognitive or affective states [38,40], such as stress, hesitation, or emotional suppression. Typical examples include brief forehead touches, lip covering, or repetitive finger tapping in high-pressure interactions. Micro-gesture

H. Chen et al. (Eds.): IJCAI 2026, CCIS 3120, pp. 17–27, 2027.
https://doi.org/10.1007/978-981-92-4529-1_2

recognition [3,27,34,41] is particularly challenging compared with general action recognition [10,11,18,23,37], as MG instances are brief, spatially subtle, and often obscured by background motion, pose-estimation noise, and large intra-class variation.

In micro-gesture recognition, previous work can be roughly grouped into skeleton-based methods [14,19] and multi-stream-based methods [1,6]. Skeleton-based methods adopt GCN, 3D-CNN [19], and transformer backbones [14] to model the subtle limb dynamics of MGs. Recent methods have further advanced micro-gesture recognition through multi-stream fusion. For instance, Chen *et al.* [1] proposed a prototype-calibrated framework that jointly models prototypes from skeleton and RGB modalities. Gu *et al.* [6] further explored large-scale multimodal fusion by integrating skeleton, RGB, depth, optical flow, and Taylor-series representations. Despite their effectiveness, these methods rely heavily on supervised annotations, while the potential of self-supervised pretraining for micro-gesture recognition remains largely unexplored.

Self-supervised video pretraining can learn transferable representations from unlabeled data, making it well-suited to scarce and imbalanced micro-gesture annotations. Recently, Wang *et al.* [34] introduced iMiGUE-3K, a large-scale micro-gesture dataset, together with MG-FM-RGB, an RGB-based self-supervised framework [32] that demonstrates the potential of self-supervised learning for micro-gesture recognition. Inspired by this work, we adopt MG-FM-RGB as a self-supervised video model for micro-gesture recognition. Specifically, the encoder is first pretrained on 120K unlabeled samples via masked video modeling and then transferred to iMiGUE for fine-tuning. Finally, we integrate this model into the previous MM-Gesture framework [6] as a complementary RGB stream for multimodal ensemble learning.

The contributions of this paper are as follows:

- We adapt a simple yet effective self-supervised RGB video framework based on SMILE masked video modeling for micro-gesture recognition. The framework leverages in-domain unlabeled videos to learn transferable RGB representations without action-level annotations.
- We integrate the self-supervised RGB stream with the existing MM-Gesture multimodal framework, enabling complementary fusion between learned RGB representations and other modality-specific cues. This combined solution achieves 74.419% top-1 accuracy on the iMiGUE test set and ranks first in Track 1 of the 4th MiGA Challenge.

2 Related Work

2.1 Micro-gesture Analysis

Micro-gesture analysis [3,27] has recently attracted growing research interest, driven by the emergence of dedicated benchmarks [7,16,21,34], advanced modeling approaches [5,17], and community competitions [2,8,20]. Micro-gesture analysis covers a range of tasks, including micro-gesture recognition [1,6,19,31], online micro-gesture detection [25,26], and behavior-based emotion understanding [43,44].

Micro-gesture recognition in the MiGA series competitions has attracted a wide range of approaches from participating teams. Li *et al.* [19] combine PoseConv3D [4] with a joint skeletal–semantic embedding loss. Chen *et al.* [1] augment PoseConv3D with a prototype-refinement module that calibrates ambiguous samples on cross-modal fused features. Huang *et al.* [15] couple Res2Net3D with an EHCT branch into a heterogeneous skeleton+RGB ensemble. Wang *et al.* [42] build a CLIP-based RGB stream with video-level distillation and inject CLIP text embeddings into the PoseConv3D skeleton stream. MM-Gesture [6] fuses six streams, including joint, limb, RGB, Taylor video, optical flow, and depth, into the previous best-performing method on iMiGUE. In contrast, Hu *et al.* [13] augment a ViT-Base RGB backbone with a global-aware importance module that focuses on fine-grained foreground regions. However, all of these methods are trained purely with supervised labels on iMiGUE, SMG, or MA-52, while the potential of self-supervised pretraining on unlabeled data remains largely unexplored.

2.2 Self-supervised Learning

Self-supervised learning [9,12,22,24,30] learns representations from unlabeled data through pretext tasks, without manual annotations. In video understanding, MAE [12] reconstructs randomly masked image patches and achieves competitive transfer to downstream tasks. VideoMAE [33] extends masked modeling to video using tube masking at high ratios. VideoMAEv2 [39] scales masked video autoencoders with dual masking, lowering pretraining cost while preserving reconstruction quality and improving transfer to downstream video understanding tasks. More recent methods reconstruct features from a frozen pretrained encoder instead of raw pixels, yielding semantically richer targets. SMILE [32] combines feature-target reconstruction with synthetic motion injection for video pretraining. Although it has demonstrated effectiveness on general action benchmarks, masked video modeling remains largely unexplored for micro-gesture recognition, where in-domain unlabeled videos are available but labeled annotations are limited. To address this gap, Wang *et al.* [34] introduced MG-FM-RGB, a SMILE-based framework pretrained on iMiGUE-3K-u and fine-tuned on iMiGUE. We therefore adopt MG-FM-RGB as the baseline model in this work.

3 Methodology

3.1 Network Architecture

As shown in Fig. 1, our solution integrates three types of models with complementary input modalities. Given an input video sample, the goal is to predict its category label $y \in \{1, \ldots, C\}$ among the C=32 iMiGUE classes. The sample is processed by K=5 streams built upon three representative backbones. For the single-stream supervised learning paradigm, we feed RGB and depth

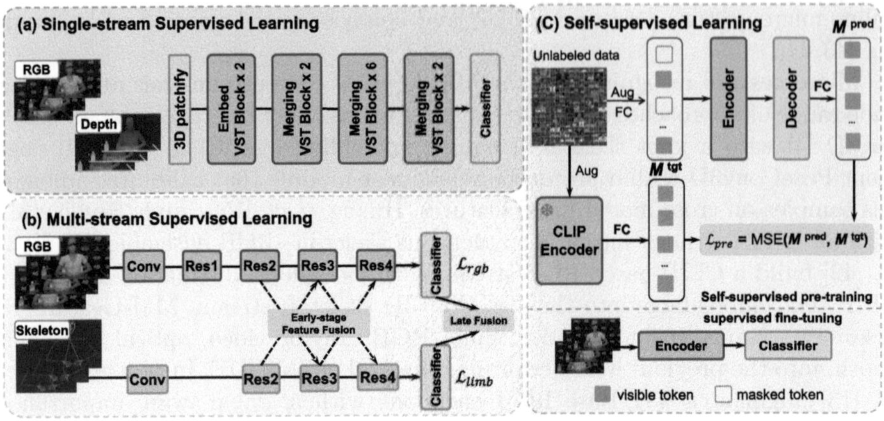

Fig. 1. Overview of the proposed multi-stream ensemble solution for micro-gesture recognition. (a) Single-stream supervised learning with Video Swin Transformer [29] using RGB and depth modalities as inputs. (b) Multi-stream supervised learning with PoseConv3D [4] to learn complementary representations from RGB and skeleton data. (c) Self-supervised learning with MG-FM-RGB [34], which is built upon SMILE [32].

inputs into Video Swin Transformer [29]. For the multi-stream supervised learning paradigm, we employ PoseConv3D [4] with skeleton and RGB inputs. For the self-supervised paradigm, we use MG-FM-RGB [34], which is pretrained on unlabeled in-domain videos. Each stream f_k outputs a softmax distribution $\boldsymbol{p}_k = (p_{k,1}, \dots, p_{k,C}) \in \mathbb{R}^C$, and the final prediction is obtained by weighted average fusion:

$$\boldsymbol{p} = \sum_{k=1}^{K} w_k \boldsymbol{p}_k, \qquad \hat{y} = \arg \max_{c \in \{1,\dots,C\}} p_c, \tag{1}$$

where $\boldsymbol{p} = (p_1, \dots, p_C)$, and $\{w_k\}_{k=1}^{K}$ denote the ensemble weights, with $w_k \geq 0$ and $\sum_{k=1}^{K} w_k = 1$.

3.2 Single-Stream Supervised Learning

Following MM-Gesture [6], we employ Video Swin Transformer [29] as the baseline model and train it with RGB and depth modalities. Video Swin Transformer extends the shifted-window self-attention mechanism of Swin Transformer [28] to 3D spatiotemporal windows, enabling the model to capture local temporal dynamics together with spatial patterns. Given an input clip, the model first splits it into 3D patch tokens. It then processes the tokens through four hierarchical stages, where regular and shifted 3D window attention are alternately applied, and patch merging layers reduce the spatial resolution between stages. We use the small variant, Swin-S, followed by global pooling and a linear classification head to predict the 32 micro-gesture categories.

The RGB stream takes the raw clip $\boldsymbol{V}_{rgb} \in \mathbb{R}^{T\times H\times W\times 3}$ as input. It is pre-trained on the Micro-Action-52 (MA-52) dataset [7] and fine-tuned on iMiGUE. The depth stream takes a depth clip $\boldsymbol{V}_{depth} \in \mathbb{R}^{T\times H\times W\times 1}$ as input. Each depth frame is generated from the corresponding RGB frame by a monocular depth estimator. Unlike the RGB stream, the depth stream is trained from scratch on iMiGUE. Let $\boldsymbol{p}_m = (p_{m,1}, \dots, p_{m,C})$ denote the softmax output of modality m. Both streams are optimized using the cross-entropy loss:

$$\mathcal{L}_m = \mathrm{CE}(\boldsymbol{p}_m, y) = -\log p_{m,y}, \quad m \in \{\mathrm{rgb}, \mathrm{depth}\}, \tag{2}$$

where $p_{m,y}$ is the probability assigned to the ground-truth class by modality m. Finally, the softmax outputs of these models are used as separate streams in the final ensemble.

3.3 Multi-stream Supervised Learning

Here, we adopt PoseConv3D [4] to model RGB and skeleton data, which has shown strong performance in previous micro-gesture recognition solutions [1,6, 45]. PoseConv3D encodes pose as heatmap volumes and fuses them with RGB at the decision level. Specifically, we first convert the 2D skeletal keypoints into heatmap representations: for each frame, every limb is drawn as a Gaussian map along the segment between its two endpoint joints, and the maps are stacked over time into a 3D limb-heatmap volume

$$\boldsymbol{H}_L \in \mathbb{R}^{T\times H\times W\times E}, \tag{3}$$

where T is the number of frames, $H\times W$ the spatial resolution, and E the number of skeletal limbs. The RGB clip $\boldsymbol{R} \in \mathbb{R}^{T\times H\times W\times 3}$ and the limb volume $\boldsymbol{H}_L$ are each encoded by a PoseConv3D backbone into a modality-specific softmax distribution $\boldsymbol{p}_m = (p_{m,1}, \dots, p_{m,C})$, $m \in \{\mathrm{R}, \mathrm{L}\}$, trained with a cross-entropy loss $\mathcal{L}_m = \mathrm{CE}(\boldsymbol{p}_m, y) = -\log p_{m,y}$.

We build two streams from these backbones. The limb stream uses the model trained on $\boldsymbol{H}_L$ alone. The skeleton+RGB stream couples the limb and RGB backbones: we fine-tune them jointly under the summed loss

$$\mathcal{L}_{R+L} = \mathcal{L}_R + \mathcal{L}_L, \tag{4}$$

and average their softmax probabilities $\boldsymbol{p}_R$ and $\boldsymbol{p}_L$ at inference,

$$\boldsymbol{p}_{R+L} = \tfrac{1}{2}\left(\boldsymbol{p}_R + \boldsymbol{p}_L\right). \tag{5}$$

Both streams are trained on iMiGUE for 32-class micro-gesture classification.

3.4 Self-supervised Learning

To learn transferable representations from large-scale unlabeled videos without manual annotations, we adopt SMILE-based MG-FM-RGB [32,34] as the self-supervised RGB stream. MG-FM-RGB is pretrained on unlabeled iMiGUE-3K [34] clips via masked video modeling and then fine-tuned on iMiGUE. Specifically, each clip is divided into N spatiotemporal tubelets, and a subset $\mathcal{M}$ is randomly masked. The visible tubelets are encoded by a transformer encoder Enc_θ, while a lightweight decoder Dec_ϕ reconstructs the masked tubelets in the feature space of a frozen CLIP encoder g_ψ, rather than in pixel space:

$$\mathcal{L}_{pre} = \frac{1}{|\mathcal{M}|} \sum_{i \in \mathcal{M}} \left\| [\text{Dec}_\phi \left(\text{Enc}_\theta(\boldsymbol{V}_{vis}) \right)]_i - [g_\psi(\boldsymbol{V})]_i \right\|_2^2 . \tag{6}$$

Here, $[\cdot]_i$ denotes the feature aligned with the i-th masked tubelet. To enhance the sparse temporal cues in micro-gesture videos, SMILE further injects synthetic motion into the clip and masks tubelets along the inserted motion trajectory. After pretraining, we fine-tune the encoder for micro-gesture classification. Specifically, we retain Enc_θ, discard the decoder, and attach a linear classification head to the mean-pooled token features. The encoder and classification head are optimized on iMiGUE using the cross-entropy loss, and the resulting softmax output is used as one stream in the final ensemble.

4 Experiments

4.1 Experimental Setup

Datasets. iMiGUE [27] is a benchmark for spontaneous micro-gesture recognition, containing 32 categories: 31 micro-gesture classes and one non-MG/background class. It provides training, validation, and test splits with 12,893, 777, and 4,562 clips and exhibits a highly long-tailed distribution, where the number of training samples per class ranges from 1 to 3,765. iMiGUE-3K [34] is a large-scale in-the-wild micro-gesture dataset collected from press-interview videos. MG-FM-RGB was pretrained on iMiGUE-3K-u, a subset of iMiGUE-3K containing 120K unlabeled clips for self-supervised pretraining. MA-52 [7] is a large-scale whole-body micro-action recognition benchmark that contains 22,422 video instances across 52 action-level categories.

Evaluation Metrics. We report top-1 classification accuracy on the iMiGUE test split, the official evaluation metric of the MiGA challenge series.

Implementation Details. Pose keypoints are extracted with OpenPose and reduced to 36 upper-body, hand, and face keypoints, from which the limb representation is formed. A per-frame depth map is estimated from the RGB video. The Video Swin RGB backbone is pretrained on MA-52 [7] and fine-tuned on iMiGUE, while the depth backbone is trained from scratch on iMiGUE. For the SMILE stream, we fine-tune the pretrained ViT-Base encoder on iMiGUE for 50 epochs with a batch size of 16, using AdamW with a peak learning rate of 5×10^{-5}, weight decay 0.05, layer-wise decay 0.75, a drop-path rate of 0.1, and a 5-epoch warmup. Each clip is sampled to 16 frames at a temporal stride of 2 and resized to 224 × 224, with mixup, cutmix, and label smoothing disabled. At inference time, we average predictions over 21 augmented views. The five streams are fused by a weighted average of their softmax probabilities, with weights tuned on the validation set.

4.2 Experimental Results

We compare our method with representative approaches from prior MiGA Track 1 submissions. Table 1 summarizes the results of HFUT-VUT [1,19], EHCT [14], MM-Gesture [6], and our solution. Our final ensemble combines PoseConv3D, Video Swin Transformer, and SMILE-based MG-FM-RGB streams, achieving 74.419% top-1 accuracy on the iMiGUE test split and ranking first in the 2026 edition. Compared with the previous best result of 73.213% from the MiGA'25 winner, our method brings a 1.206 %-point improvement.

As shown in Table 2, the self-supervised RGB stream MG-FM-RGB alone achieves 69.224% top-1 accuracy. The four-stream ensemble without the MG-FM-RGB obtains 72.591%, and incorporating the self-supervised stream MG-FM-RGB raises the accuracy to 74.419%. These results demonstrate that the self-supervised RGB representation provides complementary information to the supervised multimodal ensemble.

Table 1. Top-3 submissions of the past four MiGA Track-1 editions (2023–2026) on the iMiGUE test split, ranked by top-1 accuracy. Modalities: **J**=joint, **L**=limb, **R**=RGB, **T**=Taylor video, **F**=optical flow, **D**=depth video, **H**=heatmap.

Rank	Team	Core Methodology	Modality	Acc. (%)
MiGA'23 1st	HFUT-VUT [19]	PoseConv3D	**J** + **L**	64.12
MiGA'23 2nd	NPU-Stanford [14]	Hyperformer	**J**	63.02
MiGA'24 1st	HFUT-VUT [1]	PoseConv3D	**J** + **L** + **R**	70.254
MiGA'24 2nd	NPU-MUCIS [15]	Res2Net3D+GCN	**J** + **R**	70.188
MiGA'24 3rd	ywww11 [42]	PoseConv3D+CLIP	**J** + **R**	68.917
MiGA'25 1st	HFUT-VUT [6]	PoseConv3D+VideoSwinT	**J** + **L** + **R** + **T** + **F** + **D**	73.213
MiGA'25 2nd	awuniverse [13]	ViT-Base+GAIE	**R**	68.697
MiGA'25 3rd	Lonelysheep [45]	PoseConv3D	**J** + **L**	67.010
MiGA'26 1st	**XInsight Lab (Ours)**	PoseConv3D+VideoSwinT+MG-FM-RGB	**J** + **L** + **R** + **D**	**74.419**
MiGA'26 2nd	YUV	-	-	74.243
MiGA'26 3rd	xd	-	-	71.372
MiGA'26 4th	AIM [46]	Decoupled ST-CNN + PoseC3D + Swin3D + R(2+1)D	**J** + **H** + **R**	68.128

Finally, we analyze the multimodal ensemble strategy. As shown in Table 2, we construct an ensemble using multiple models, including PoseConv3D [4] and Video Swin Transformer [29] with the modalities adopted in MM-Gesture [6], and MG-FM-RGB [34], a SMILE-based self-supervised RGB model that we fine-tune on iMiGUE. By incorporating this self-supervised stream into the existing ensemble, the top-1 accuracy improves from 72.591% to 74.419%. These results indicate that the proposed RGB-based self-supervised baseline provides complementary representations to the existing RGB, depth, and limb-based streams, thereby further enhancing the overall ensemble performance.

Table 2. Ensemble accuracy on iMiGUE test set under different models: PoseC3D [4], Video Swin Transformer [29], MG-FM-RGB* denotes our implementation results using official code.

Model	PoseC3D-L	PoseC3D-LR	VideoSwinT	VideoSwinT	MG-FM-RGB*	Acc. (%)
Modality	Limb	Limb + RGB	RGB	Depth	RGB	
Pre-training	iMiGUE	iMiGUE	MA-52	N/A	iMiGUE-3K-u	
Fine-tuning	iMiGUE	iMiGUE	iMiGUE	iMiGUE	iMiGUE	
Single	✓					64.686
		✓				68.917
			✓			66.615
				✓		65.212
					✓	69.224
Ensemble	✓	✓	✓	✓		72.591
	✓	✓	✓	✓	✓	**74.419**

5 Conclusion

In this paper, we presented a simple multi-stream ensemble for micro-gesture recognition. The streams are independently trained or fine-tuned under distinct training regimes, ranging from supervised pretraining for skeleton and video streams to self-supervised masked video modeling on iMiGUE-3K-u; their predictions are fused by a weighted softmax average. On the iMiGUE test split, the strongest single stream reaches a top-1 accuracy of 69.224%, while the full ensemble reaches 74.419%, ranking first in Track 1 of the 4th MiGA Challenge at IJCAI 2026.

In future work, we plan to design micro-gesture-specific self-supervised pre-training with motion magnification [35,36] to better capture the short-duration and low-amplitude motion of micro-gestures, and to explore semantic priors from multimodal large language models for ambiguous and long-tailed categories.

Acknowledgments. This work was supported by Anhui Provincial Natural Science Foundation (2408085J040), National Key R&D Program of China (2024YFB3311600), National Natural Science Foundation of China (62272144, 72188101), the Major Project of Anhui Provincial Science and Technology Breakthrough Program (202423k09020001), and the New Cornerstone Science Foundation through the XPLORER PRIZE.

References

1. Chen, G., et al.: Prototype learning for micro-gesture classification. arXiv preprint arXiv:2408.03097 (2024)
2. Chen, H., Schuller, B.W., Adeli, E., Zhao, G.: The 3rd challenge on human behavior analysis for emotion understanding (MiGA) 2025: From recognition to emotion understanding (2025)

3. Chen, H., Shi, H., Liu, X., Li, X., Zhao, G.: SMG: a micro-gesture dataset towards spontaneous body gestures for emotional stress state analysis. Int. J. Comput. Vision **131**(6), 1346–1366 (2023)
4. Duan, H., Zhao, Y., Chen, K., Lin, D., Dai, B.: Revisiting skeleton-based action recognition. In: Proceedings of the IEEE/CVF Conference on Computer Vision and Pattern Recognition, pp. 2969–2978 (2022)
5. Gu, J., et al.: Motion matters: motion-guided modulation network for skeleton-based micro-action recognition. In: Proceedings of the 33rd ACM International Conference on Multimedia. pp. 5461–5470 (2025)
6. Gu, J., Wang, F., Li, K., Wei, Y., Wu, Z., Guo, D.: MM-Gesture: towards precise micro-gesture recognition through multimodal fusion. arXiv preprint arXiv:2507.08344 (2025)
7. Guo, D., Li, K., Hu, B., Zhang, Y., Wang, M.: Benchmarking micro-action recognition: dataset, methods, and applications. IEEE Trans. Circuits Syst. Video Technol. **34**(7), 6238–6252 (2024)
8. Guo, D., et al.: MAC 2024: micro-action analysis grand challenge. In: Proceedings of the 32nd ACM International Conference on Multimedia. pp. 11304–11305 (2024)
9. Guo, T., Liu, H., Chen, Z., Liu, M., Wang, T., Ding, R.: Contrastive learning from extremely augmented skeleton sequences for self-supervised action recognition. In: Proceedings of the AAAI Conference on Artificial Intelligence, vol. 36, pp. 762–770 (2022)
10. Hao, Y., et al.: Attention in attention: modeling context correlation for efficient video classification. IEEE Trans. Circuits Syst. Video Technol. **32**(10), 7120–7132 (2022)
11. Hao, Y., Zhang, H., Ngo, C.W., He, X.: Group contextualization for video recognition. In: Proceedings of the IEEE/CVF Conference on Computer Vision and Pattern Recognition, pp. 928–938 (2022)
12. He, K., Chen, X., Xie, S., Li, Y., Dollár, P., Girshick, R.: Masked autoencoders are scalable vision learners. In: Proceedings of the IEEE/CVF Conference on Computer Vision and Pattern Recognition, pp. 16000–16009 (2022)
13. Hu, X., Pu, C., Li, Y., Xu, Y., Xie, K., Miao, Q.: Enhancing micro-gesture classification via global-aware importance estimation in vision transformer. In: MiGA@ IJCAI (2025)
14. Huang, H., Guo, X., Peng, W., Xia, Z.: Micro-gesture classification based on ensemble hypergraph-convolution transformer. In: MiGA@ IJCAI (2023)
15. Huang, H., Wang, Y., Linghu, K., Xia, Z.: Multi-modal micro-gesture classification via multi-scale heterogeneous ensemble network. In: MiGA@ IJCAI (2024)
16. Li, K., Gu, J., Wang, F., Wu, Z., Fan, H., Guo, D.: MA-Bench: towards fine-grained micro-action understanding. In: Proceedings of the IEEE/CVF Conference on Computer Vision and Pattern Recognition, pp. 20118–20128 (2026)
17. Li, K., et al.: Prototypical calibrating ambiguous samples for micro-action recognition. In: Proceedings of the AAAI Conference on Artificial Intelligence, vol. 39, pp. 4815–4823 (2025)
18. Li, K., Guo, D., Chen, G., Liu, F., Wang, M.: Data augmentation for human behavior analysis in multi-person conversations. In: Proceedings of the 31st ACM International Conference on Multimedia, pp. 9516–9520 (2023)
19. Li, K., Guo, D., Chen, G., Peng, X., Wang, M.: Joint skeletal and semantic embedding loss for micro-gesture classification (2023). arXiv:2307.10624 arXiv preprint
20. Li, K., et al.: MAC 2025: the 2nd micro-action analysis grand challenge. In: Proceedings of the 33rd ACM International Conference on Multimedia, pp. 14216–14221 (2025)

21. Li, K., et al.: MMAD: multi-label micro-action detection in videos. In: Proceedings of the IEEE/CVF International Conference on Computer Vision, pp. 13225–13236 (2025)
22. Li, L., Wang, M., Ni, B., Wang, H., Yang, J., Zhang, W.: 3D human action representation learning via cross-view consistency pursuit. In: Proceedings of the IEEE/CVF Conference on Computer Vision and Pattern Recognition, pp. 4741–4750 (2021)
23. Li, Q., et al.: Data-efficient masked video modeling for self-supervised action recognition. In: Proceedings of the 31st ACM International Conference on Multimedia, pp. 2723–2733 (2023)
24. Lin, L., Song, S., Yang, W., Liu, J.: MS2L: multi-task self-supervised learning for skeleton based action recognition. In: Proceedings of the 28th ACM International Conference on Multimedia, pp. 2490–2498 (2020)
25. Liu, P., Li, K., Wang, F., Wei, Y., She, J., Guo, D.: Online micro-gesture recognition using data augmentation and spatial-temporal attention (2025). arXiv:2507.09512 arXiv preprint
26. Liu, P., et al.: Micro-gesture online recognition using learnable query points (2024). arXiv:2407.04490 arXiv preprint
27. Liu, X., Shi, H., Chen, H., Yu, Z., Li, X., Zhao, G.: iMiGUE: an identity-free video dataset for micro-gesture understanding and emotion analysis. In: Proceedings of the IEEE/CVF Conference on Computer Vision and Pattern Recognition, pp. 10631–10642 (2021)
28. Liu, Z., et al.: Swin Transformer: hierarchical vision transformer using shifted windows. In: Proceedings of the IEEE/CVF International Conference on Computer Vision, pp. 10012–10022 (2021)
29. Liu, Z., et al.: Video swin transformer. In: Proceedings of the IEEE/CVF Conference on Computer Vision and Pattern Recognition, pp. 3202–3211 (2022)
30. Mao, Y., Deng, J., Zhou, W., Fang, Y., Ouyang, W., Li, H.: Masked motion predictors are strong 3D action representation learners. In: Proceedings of the IEEE/CVF International Conference on Computer Vision, pp. 10181–10191 (2023)
31. Shang, T., Hao, Y., Pei, M., Li, K., Ben, H., Wang, S.: Cross-modal feature enhancement and contrastive alignment for micro-gesture recognition. In: Chinese Conference on Pattern Recognition and Computer Vision (PRCV), pp. 203–217. Springer (2025)
32. Thoker, F.M., Jiang, L., Zhao, C., Ghanem, B.: SMILE: infusing spatial and motion semantics in masked video learning. In: Proceedings of the Computer Vision and Pattern Recognition Conference, pp. 8438–8449 (2025)
33. Tong, Z., Song, Y., Wang, J., Wang, L.: VideoMAE: masked autoencoders are data-efficient learners for self-supervised video pre-training. Adv. Neural. Inf. Process. Syst. **35**, 10078–10093 (2022)
34. Wang, C., Chen, H., Wei, H., Yang, Y., Chen, Y., Zhao, G.: iMiGUE-3K: a large-scale benchmark for micro-gesture analysis with self-supervised learning. arXiv preprint arXiv:2605.17179 (2026)
35. Wang, F., Guo, D., Li, K., Wang, M.: EulerMormer: robust Eulerian motion magnification via dynamic filtering within transformer. In: Proceedings of the AAAI Conference on Artificial Intelligence, vol. 38, pp. 5345–5353 (2024)
36. Wang, F., Guo, D., Li, K., Zhong, Z., Wang, M.: Frequency decoupling for motion magnification via multi-level isomorphic architecture. In: Proceedings of the IEEE/CVF Conference on Computer Vision and Pattern Recognition, pp. 18984–18994 (2024)

37. Wang, F., et al.: Exploiting ensemble learning for cross-view isolated sign language recognition. In: Companion Proceedings of the ACM on Web Conference 2025, pp. 2453–2457 (2025)
38. Wang, F., et al.: XInsight: integrative stage-consistent psychological counseling support agents for digital well-being. In: Proceedings of the ACM Web Conference 2026, pp. 9297–9308 (2026)
39. Wang, L., et al.: VideoMAE V2: scaling video masked autoencoders with dual masking. In: Proceedings of the IEEE/CVF Conference on Computer Vision and Pattern Recognition, pp. 14549–14560 (2023)
40. Wang, R., Li, K., Tong, A., Xu, J., Guo, D., Wang, M.: Gait emotion recognition via uncertainty-oriented class discriminative learning. IEEE Trans. Affect. Comput., 1–14 (2026)
41. Wang, T., et al.: Micro-gesture recognition: a comprehensive survey of datasets, methods, and challenges. Mach. Intell. Res. **23**(2), 308–330 (2026)
42. Wang, Y., Dong, Z., Li, P., Liu, Y.: A multimodal micro-gesture classification model based on clip. In: MiGA@ IJCAI (2024)
43. Wang, Y., Liu, H., Xu, T., Shi, C., Xing, H.: Weak to Strong: VLM-based pseudo-labeling as a weakly supervised training strategy in multimodal video-based hidden emotion understanding tasks (2026). arXiv:2602.08057 arXiv preprint
44. Xia, Z., Huang, H., Chen, H., Feng, X., Zhao, G.: Hybrid-supervised hypergraph-enhanced transformer for micro-gesture based emotion recognition. IEEE Trans. Affect. Comput., (2025)
45. Xu, H., Cheng, L., Wang, Y., Tang, S., Zhong, Z.: Towards fine-grained emotion understanding via skeleton-based micro-gesture recognition (2025). arXiv:2506.12848 arXiv preprint
46. Zhang, H., Zhang, H., Liu, P., Zhang, Y., Xue, W., Hao, Y.: A multi-modal framework with cross-subject pseudo-labeling and semantic alignment for micro-gesture recognition. arXiv preprint arXiv:2606.13030 (2026)

SwinT-Enhanced Class-Wise Fusion for Micro-gesture Classification

Taorui Wang[1,2], Hui Ma[2], Wei Xia[2], Jiayu Zhang[2], Jiajian Huang[2], and Yong Xu[1](✉)

[1] Harbin Institute of Technology, Shenzhen, China
wangtaorui@stu.hit.edu.cn, laterfall@hit.edu.cn
[2] Great Bay University, China, China

Abstract. Micro-gesture classification depends on short, low-amplitude, and class-dependent cues, making global ensemble weights brittle when many frozen experts are available. We present CLOG-Fusion, a 416-parameter class-wise log-opinion fusion head for The 4th MiGA-IJCAI Challenge Track 1: Micro-gesture Classification, an official Kaggle competition. The implementation uses a SwinT-enhanced expert pool built from iMiGUE-finetuned MM-Gesture SwinT backbones plus auxiliary motion, MA-52-pretrained, and PoseC3D skeleton experts. The final YUV submission achieved a 0.74243 official-test leaderboard score and ranked second on the completed leaderboard.

Keywords: Micro-gesture recognition · Multimodal fusion · Ensemble learning · iMiGUE

1 Introduction

Micro-gestures (MGs) are spontaneous, fine-grained movements such as subtle hand, face, or upper-body adjustments. Compared with conventional action recognition, MG classification is more sensitive to short temporal evidence, small spatial regions, and long-tail labels. iMiGUE introduced identity-free micro-gesture understanding [9], and iMiGUE-3K expands this line to larger-scale 32-class analysis [14]. Recent survey work identifies multimodal fusion and robust recognition under limited data as central challenges [15].

Our competition observation is that expert reliability is class dependent. In the SwinT-enhanced pool, visual experts handle appearance classes; motion and skeleton experts cover displacement or pose cues. Global averaging or a single validation-tuned ensemble weight cannot express this per-category behavior. Instead of training another heavy recognizer, we place a compact fusion head on top of frozen expert logits.

The resulting method, CLOG-Fusion, is a constrained log-opinion ensemble. It keeps the agreement-sharpening behavior of product-style opinion pooling, but learns a separate expert simplex for each target class. This design lets the

H. Chen et al. (Eds.): IJCAI 2026, CCIS 3120, pp. 28–34, 2027.
https://doi.org/10.1007/978-981-92-4529-1_3

paper make a narrow, auditable claim: class-wise reliability can be learned with a small number of parameters over a SwinT-enhanced expert pool.

The main contributions are:

- We present a lightweight class-wise log-opinion fusion head over a SwinT-enhanced expert pool with visual, motion-derived, MA-52-pretrained [4], and PoseC3D skeleton experts.
- We quantify expert complementarity with single-expert validation metrics, fusion ablations, and a learned class-by-expert weight visualization.
- We report the final competition outcome: the YUV submission ranked second on the completed leaderboard.

2 Related Work

Micro-gesture Recognition. iMiGUE and iMiGUE-3K define identity-free and larger 32-class MG benchmarks [9,14]. MA-52 is a neighboring 52-class micro-action source for subtle-motion pretraining [4]. Prior challenge systems, including MM-Gesture, show that strong submissions benefit from complementary visual, motion, and skeleton cues [3]. Neighboring subtle-visual-computing work, including SVC 2025 multimodal deception detection and broader deception-detection surveys, also emphasizes multimodal cues and cross-domain robustness [8,17]. We use this line of work to motivate expert diversity, while the contribution here is last-stage fusion over a SwinT-enhanced expert pool rather than a new backbone.

Video and Skeleton Experts. Video Swin Transformer provides the SwinT-style visual backbone used by our RGB, transformed-video, and motion-derived experts [10]. PoseC3D converts skeleton sequences into 3D heatmap volumes for joint-motion evidence [2]. Our method does not modify these backbones; it asks how to combine stored predictions when validation data are limited.

Multimodal Late Fusion. Late fusion reuses specialists without retraining all backbones. Classifier-combination and early/late-fusion studies motivate averaging and validation-tuned weights [1,7,12]. Stacking and mixture-of-experts models are more flexible, but can overfit a 777-sample validation split [6,16]. Product-of-experts pooling sharpens agreement [5]; feature conditioning such as FiLM requires feature-level retraining [11]. CLOG-Fusion keeps log-opinion pooling while restricting learnable parameters to class-wise expert weights and a class bias.

3 Methodology

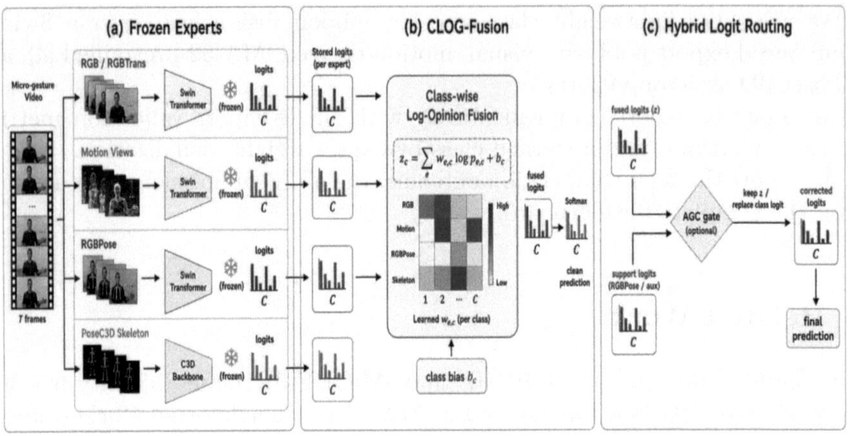

Fig. 1. Pipeline of the proposed system: frozen experts, class-wise log-opinion fusion, and final submission assembly.

3.1 Frozen Expert Generation

As shown in Fig. 1, the system begins with independently trained experts. The main appearance experts are SwinT RGB and RGBTrans video baselines implemented in the MM-Gesture/MMAction2 code path [3,10]. In the released submission code, the submitted visual logits come from MM-Gesture SwinT backbone weights that had already been fine-tuned on iMiGUE, rather than from generic ImageNet-only initialization. Thus, "frozen" means frozen during the fusion stage: several logit sources are challenge-trained or MA-52/iMiGUE-fine-tuned experts before their predictions are stored. Complementary experts include Taylor videos, MemFlow and motion-region crops, MA-52-pretrained RGBTrans variants [4], and PoseC3D joint/limb skeleton branches [2]. Heavy training and inference are completed before fusion; the fusion stage consumes stored probabilities for 12,893 train clips, 777 validation clips, and 4,562 official-test rows.

Taylor-S and MemFlow-S have the same rounded top-1 count, but they are distinct experts with different top-5 and NLL behavior in the validation audit. The table is a reliability inventory: weaker single experts can still help when their errors are class-specific and partially independent from the appearance baseline.

MRRF-crop is a deterministic motion-region preprocessing expert rather than a separate detector. For each video, we accumulate grayscale adjacent-frame differences, blur the accumulated motion map, threshold it by the larger of the 82nd positive-motion percentile and a mean-plus-standard-deviation rule, and close the mask with a 9×9 kernel. The resulting bounding box is expanded

Table 1. Frozen experts used by the main 12-expert CLOG-Fusion run.

Expert	Modality/input	Backbone	Val top-1
RGB-S	RGB video	SwinT-S	73.49
RGBTrans-S	transformed RGB	SwinT-S	76.06
Taylor-S	Taylor motion video	SwinT-S	69.11
MemFlow-S	motion video	SwinT-S	69.11
MRRF-crop	motion-region crop	SwinT-S	73.36
RGB-B	RGB video	SwinT-B	75.03
Flow-B	optical-flow video	SwinT-B	68.85
Taylor-B	Taylor motion video	SwinT-B	68.34
MA52-final	MA-52-pretrained RGBTrans	SwinT-S	73.62
MA52-early	MA-52-pretrained RGBTrans	SwinT-S	70.91
PoseC3D-J	22-keypoint joint skeleton	PoseC3D	62.81
PoseC3D-L	22-keypoint limb skeleton	PoseC3D	60.10

by a 0.35 margin, forced to cover at least 10% of the frame area, made approximately square, resized to 256×256, and then passed through the same SwinT-S video classifier. In the mask-crop variant, the crop is blended with a soft motion mask over a blurred background so that low-motion context is retained while the moving region is emphasized.

3.2 Class-Wise Log-Opinion Fusion

For a clip x, expert e outputs a probability vector $p_e(x) \in [0,1]^C$ over C micro-gesture classes. CLOG-Fusion learns an expert simplex separately for each class:

$$w_{e,c} = \mathrm{softmax}_e(\alpha_{e,c}), \qquad z_c(x) = \sum_e w_{e,c} \log\left(\max(p_{e,c}(x), \epsilon)\right) + b_c. \quad (1)$$

The final distribution is $q(y = c \mid x) = \mathrm{softmax}_c(z_c(x))$. The head is fitted on frozen logits with a smoothed cross-entropy objective and mild regularization toward non-degenerate class-wise expert mixtures.

For the main $E = 12$, $C = 32$ setting, the head contains $EC + C = 416$ trainable scalar parameters. We avoid input-conditioned routing because 777 validation clips make high-capacity stackers brittle; CLOG-Fusion instead forms a class-level geometric opinion pool that reinforces agreement and downweights unreliable expert-class pairs.

4 Experiments

Experimental Setup. We follow the train/validation/test split used in The 4th MiGA-IJCAI Challenge Track 1: Micro-gesture Classification [13]. Training labels fit the fusion head, validation data selects fusion hyperparameters, and official-test predictions are submitted without label access. We optimize CLOG-Fusion with AdamW using learning rate 0.04, batch size 2048, up to 900 epochs, and early stopping patience 140. The loss uses label smoothing 0.03, uniform-simplex weight regularization, bias L_2, and entropy regularization 0.001; the regularization grid is $\lambda_w \in \{0.001, 0.003, 0.01, 0.03\}$ and $\lambda_b \in \{0, 0.001, 0.003\}$. The submission verifier checks the `Id,Target` format, 4,562 rows, unique IDs, and valid target ranges. Because official-test labels are hidden, Table 2 is validation evidence for method selection rather than an official-test ablation. No official-test labels are used to fit the fusion head, and validation metrics and final leaderboard rank are reported separately to avoid interpreting leaderboard provenance as per-component test-set evidence.

4.1 Fusion Ablation

Table 2. Fusion ablations. Validation metrics are used for method selection; final leaderboard rank is reported separately.

System row	Experts	Val top-1	Val NLL
RGB/RGBTrans average	2	75.03	–
RGB/RGBTrans global weighted	2	76.45	–
12-expert probability average	12	76.19	0.794
12-expert class-wise probability fusion	12	76.45	0.780
CLOG-Fusion without MA52	10	77.48	0.769
CLOG-Fusion with one MA52 expert	11	77.35	0.768
Core CLOG-Fusion	12	77.09	0.766

Table 2 summarizes validation evidence for the SwinT-enhanced method. The 12-expert average is the multi-expert reference, and CLOG-Fusion reduces validation NLL through class-specific log-space weights. Adding MA52 experts [4] improves NLL but not top-1, so we treat it as complementary calibration evidence rather than top-1 evidence. The class-wise probability-fusion row uses the same $EC + C = 416$ scalar budget and the same training grid as CLOG-Fusion, but replaces $\sum_e w_{e,c} \log p_{e,c}$ with $\log(\sum_e w_{e,c} p_{e,c})$. Its best validation-NLL candidate remains behind the log-opinion head, suggesting that the gain is not only from class-wise specialization.

4.2 Per-Class Analysis

Figure 2 shows the class-wise simplex learned by CLOG-Fusion. The pattern is soft, as expected from entropy and uniform-simplex regularizers, but it is not global: different classes select different largest-weight experts. Validation recall also shows the long tail. Class 30 has 193 validation samples and 94.82% recall, while classes 1 and 20 have 25/24.00% and 13/30.77%; three singleton classes are missed. The analysis supports class-dependent reliability while showing rare classes as the main limitation.

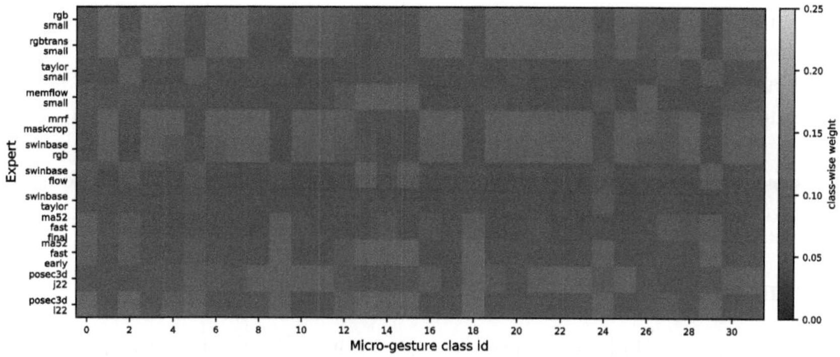

Fig. 2. Learned class-wise expert weights for the main CLOG-Fusion run. The weights are regularized and therefore not one-hot, but the preferred expert varies by class, with MRRF-crop the largest-weight expert for 16 of 32 classes and the remaining classes split across RGB, MA52, PoseC3D, and motion experts.

Table 3. Final Kaggle leaderboard for The 4th MiGA-IJCAI Challenge Track 1. These official-test scores are reported as competition standings, not as validation evidence for model selection.

Rank	Team	Leaderboard score
1	XInsight Lab	0.74660
2	YUV (ours)	0.74243
3	xd	0.71372
4	AIM	0.68128

4.3 Competition Result

The completed competition is listed on Kaggle as The 4th MiGA-IJCAI Challenge Track 1: Micro-gesture Classification [13]. Table 3 reports the final leaderboard standings directly. The YUV submission achieved 0.74243 and ranked second, behind XInsight Lab at 0.74660. These leaderboard scores are competition provenance, while Tables 1 and 2 remain validation evidence for the class-wise fusion method itself.

Limitations. The official leaderboard evaluates the final submitted file, but the hidden labels prevent a clean official-test decomposition of each fusion component. Therefore, the validation ablations should be read as model-selection evidence, not as unbiased estimates of the official-test gains. Rare classes also remain difficult: several validation classes have very low support, and singleton classes cannot be reliably recovered by a compact fusion head alone.

5 Conclusion

We presented CLOG-Fusion, a compact class-wise log-opinion fusion method with per-class reliability over a SwinT-enhanced expert pool spanning visual, motion, and skeleton recognizers. The final YUV submission achieved 0.74243 and ranked second on the completed leaderboard. The main remaining limitation is rare-class recognition and the validation-only nature of component ablations under hidden official-test labels.

Disclosure of Interests. The author has no competing interests to declare that are relevant to the content of this article.

References

1. Baltrušaitis, T., Ahuja, C., Morency, L.P.: Multimodal machine learning: a survey and taxonomy. IEEE Trans. Pattern Anal. Mach. Intell. **41**(2), 423–443 (2019)
2. Duan, H., Zhao, Y., Chen, K., Lin, D., Dai, B.: Revisiting skeleton-based action recognition. In: CVPR (2022)
3. Gu, J., Wang, F., Li, K., Wei, Y., Wu, Z., Guo, D.: MM-Gesture: towards precise micro-gesture recognition through multimodal fusion. arXiv preprint arXiv:2507.08344 (2025)
4. Guo, D., Li, K., Hu, B., Zhang, Y., Wang, M.: Benchmarking micro-action recognition: dataset, methods, and applications. IEEE Trans. Circuits Syst. Video Technol. **34**(7), 6238–6252 (2024)
5. Hinton, G.E.: Training products of experts by minimizing contrastive divergence. Neural Comput. **14**(8), 1771–1800 (2002)
6. Jacobs, R.A., Jordan, M.I., Nowlan, S.J., Hinton, G.E.: Adaptive mixtures of local experts. Neural Comput. **3**(1), 79–87 (1991)
7. Kittler, J., Hatef, M., Duin, R.P.W., Matas, J.: On combining classifiers. IEEE Trans. Pattern Anal. Mach. Intell. **20**(3), 226–239 (1998)
8. Lin, X., et al.: SVC 2025: the first multimodal deception detection challenge. In: Proceedings of the 1st International Workshop & Challenge on Subtle Visual Computing, pp. 59–64 (2025)
9. Liu, X., Shi, H., Chen, H., Yu, Z., Li, X., Zhao, G.: iMiGUE: an identity-free video dataset for micro-gesture understanding and emotion analysis. In: CVPR (2021)
10. Liu, Z., Ning, J., Cao, Y., Wei, Y., Zhang, Z., Lin, S., Hu, H.: Video Swin transformer. In: CVPR (2022)
11. Perez, E., Strub, F., de Vries, H., Dumoulin, V., Courville, A.: FiLM: visual reasoning with a general conditioning layer. In: AAAI (2018)
12. Snoek, C.G.M., Worring, M., Smeulders, A.W.M.: Early versus late fusion in semantic video analysis. In: ACM Multimedia, pp. 399–402 (2005)
13. Wang, C.: The 4th MiGA-IJCAI challenge track 1: Micro-gesture classification. https://www.kaggle.com/competitions/the-4th-ei-mi-ga-ijcai-challenge-track-1 (2026). kaggle
14. Wang, C., Chen, H., Wei, H., Yang, Y., Chen, Y., Zhao, G.: iMiGUE-3K: a large-scale benchmark for micro-gesture analysis with self-supervised learning. arXiv preprint arXiv:2605.17179 (2026)
15. Wang, T., et al.: Micro-gesture recognition: a comprehensive survey of datasets, methods, and challenges. Mach. Intell. Res. **23**(2), 308–330 (2026). https://doi.org/10.1007/s11633-025-1629-x
16. Wolpert, D.H.: Stacked generalization. Neural Netw. **5**(2), 241–259 (1992)
17. Zhang, J., et al.: Multimodal deception detection: a survey. Mach. Intell. Res. **23**(2), 284–307 (2026)

Skeleton-Based Micro-gesture Recognition via Information Bottleneck Memory and Joint-Bone Mutual Alignment

Kunyang Wu[1] and Yunan Li[1,2,3](✉)

[1] School of Computer Science and Technology, Xidian University, China, China
[2] Xi'an Key Laboratory of Big Data and Intelligent Vision, China, China
[3] Key Laboratory of Smart Human-Computer Interaction and Wearable Technology of Shaanxi Province, China, China
yunanli@xidian.edu.cn

Abstract. Micro-gesture classification is challenging due to subtle motion patterns, short durations, and large intra-class variations. Although RGB and skeleton modalities are commonly used, discriminative skeleton representations remain insufficiently explored. Specifically, skeleton features may contain task-irrelevant variations caused by performer-specific styles, keypoint noise, and indirect visual disturbances. Meanwhile, joint and bone streams are often fused without explicit semantic coordination, limiting their complementary potential.To address these issues, we propose a skeleton representation learning framework for micro-gesture recognition. Specifically, IB-Memory suppresses task-irrelevant variations and distills category-shared predictive semantics. In addition, JBMA aligns the prediction distributions of joint and bone streams via bi-directional KL divergence, enhancing their semantic complementarity. Experiments on the MiGA 2026 Challenge show that our method achieves 3rd place, validating the effectiveness of skeleton feature purification and joint-bone mutual alignment.

Keywords: Micro-gesture classification · Skeleton representation learning · Information Bottleneck · Memory Network

1 Introduction

Micro-gestures [3,11,14] are spontaneous, fine-grained movements that arise involuntarily under stress or emotional fluctuation, typically manifesting as subtle physical changes such as facial micro-expressions, shoulder and neck contractions, and subtle hand movements. These gestures encode rich non-verbal emotional cues, making them highly valuable for implicit emotion perception, intelligent human-computer interaction, and mental health assessment. Unlike macro-scale body movements with explicit communicative intent, micro-gestures exhibit

H. Chen et al. (Eds.): IJCAI 2026, CCIS 3120, pp. 35–45, 2027.
https://doi.org/10.1007/978-981-92-4529-1_4

low movement amplitudes, high inter-class similarity, and pronounced individual differences, which collectively pose significant challenges for fine-grained classification.

Mainstream micro-gesture recognition methods often adopt multimodal frameworks that fuse RGB and skeletal data [2,6]. RGB provides rich appearance and contextual cues but is sensitive to illumination, background clutter, and subject appearance. Skeleton data offers a compact motion representation with stronger robustness to visual interference, yet existing skeleton models still suffer from task-irrelevant variations and insufficient coordination between joint and bone streams. In particular, simple late fusion cannot fully exploit the semantic complementarity between these two skeleton representations.

To overcome the aforementioned bottlenecks, we focus on optimizing skeletal representations built upon the SkateFormer [5] baseline. Our main contributions are summarized as follows:

- **Information Bottleneck Memory Model (IB-Memory):** We introduce IB-Memory as a task-specific purification mechanism for micro-gesture recognition. While previously used in macro-level tasks, we demonstrate its unique effectiveness in filtering out performer-specific jitters and subtle habit variations inherent to micro-gestures. By maintaining class-specific prototypes, it forces the backbone to distill robust motion semantics, serving as a vital prerequisite for multi-stream alignment.
- **Joint-Bone Mutual Alignment (JBMA):** We introduce JBMA to align joint and bone streams at the prediction-distribution level with bi-directional KL consistency, while preserving their complementary motion cues through feature-level concatenation.

2 Related Works

2.1 Micro-gesture Classification

Existing micro-gesture classification methods mainly rely on RGB videos, skeleton sequences, or multimodal fusion. RGB-based methods exploit appearance cues and spatiotemporal context, but they are sensitive to background clutter, illumination changes, and subject appearance variations. Skeleton-based methods provide a compact geometric representation of human motion and are more robust to visual interference. Recent MiGA solutions often combine RGB, skeleton, optical flow, or model ensembles to improve recognition accuracy [2,6]. However, most of them focus on modality-level prediction fusion, while the internal representation learning of the skeleton modality remains underexplored. In particular, the semantic relationship between joint and bone representations is usually modeled by simple late fusion, which limits their complementary potential.

2.2 Skeleton-based Action Recognition

Skeleton-based action recognition has been widely studied due to the robustness and compactness of skeletal data. Early methods such as ST-GCN [17] model human joints with predefined graph structures, but their receptive fields are limited by fixed physical connections. Recent Transformer-based methods overcome this limitation by using self-attention to capture long-range spatiotemporal dependencies. SkateFormer [5] further improves skeleton representation learning by modeling spatial and temporal patterns in a decoupled manner, making it a strong baseline for fine-grained motion recognition. Nevertheless, existing skeleton models usually treat joint and bone streams as separate inputs and combine them only at the late prediction stage. This motivates us to explicitly align joint and bone semantics before fusion.

2.3 Information Bottleneck and Memory Networks

The Information Bottleneck (IB) principle [1,13] aims to preserve task-relevant predictive information while suppressing redundant factors in feature representations. Memory networks [16] provide an explicit mechanism to store and retrieve representative features from samples. Recent work combines IB theory with memory networks to distill robust gesture representations by weakening task-irrelevant variations [9]. Inspired by this idea, we adapt the IB-Memory mechanism to skeleton-based micro-gesture recognition. Although originally designed for macro-level RGB-D gestures, we empirically reveal its critical role in fine-grained tasks: micro-gestures are heavily plagued by task-irrelevant variations, such as individual nervous habits or subtle tracking jitters. By applying IB-Memory to explicitly decouple and purify joint and bone skeleton streams, we provide a stable, noise-invariant foundation that is essential for our subsequent joint-bone mutual alignment.

3 Methodology

3.1 Overall Network Framework

We propose a multimodal framework for micro-gesture classification, consisting of a skeleton branch, an RGB branch, and a lightweight decision-level fusion stage. The skeleton branch serves as the main optimization target for capturing fine-grained motion structures, while the RGB branch provides complementary appearance and spatiotemporal cues. The predictions from both modalities are finally integrated by temperature-calibrated fusion with a validation-selected global weight.

In the skeleton branch, joint and bone representations are used as two inputs. The joint stream mainly describes keypoint locations and motion trajectories, while the bone stream captures relative structures and local motion directions between adjacent joints. Since both streams are derived from the same skeleton sequence, they have a natural basis for category-level semantic consistency. Based on this homogeneous relationship, JBMA imposes bi-directional KL consistency on their prediction distributions and uses feature-level concatenation to preserve complementary expressions, thereby enhancing skeleton discrimination.

Considering that RGB and Skeleton are heterogeneous modalities, they differ in information sources and noise patterns: RGB focuses more on appearance textures and visual context, whereas Skeleton emphasizes human structure and motion trajectories. Therefore, instead of introducing complex feature-level interaction or strong semantic alignment between RGB and Skeleton, we adopt a more stable decision-level fusion strategy to preserve the discriminative advantages of both modalities (Fig. 1).

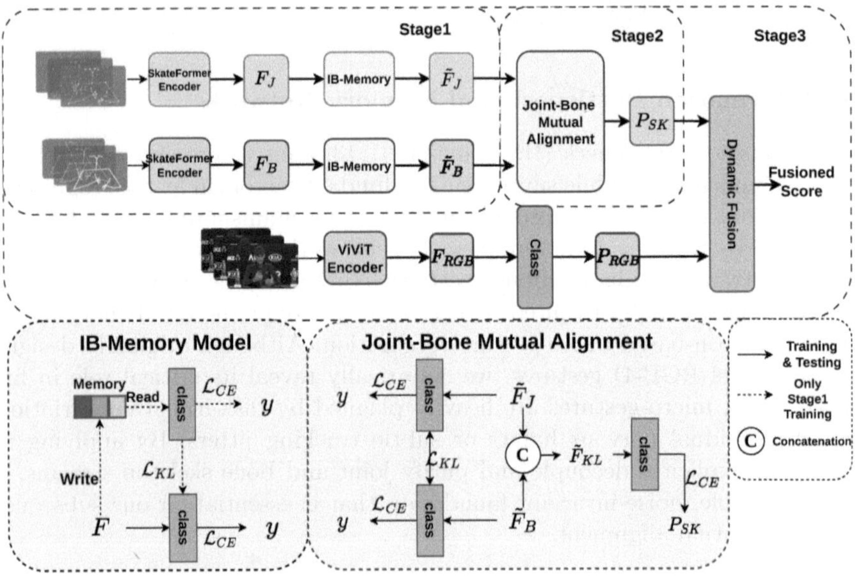

Fig. 1. Overall framework of the proposed method. IB-Memory purifies single-stream skeleton features, JBMA aligns and fuses joint-bone representations, and calibrated fusion integrates RGB and skeleton predictions. Solid arrows denote training-and-inference paths, dashed arrows denote Stage-1-only paths, and C denotes concatenation.

3.2 Information Bottleneck Memory Model

Given the deep features $v \in \{F_J, F_B\}$ extracted by the single-stream skeleton backbone, IB-Memory maintains a class-specific memory bank for each gesture category. For a sample with label y, the corresponding memory bank is denoted as $M^y \in \mathbb{R}^{K \times C}$, where K is the number of memory slots for each class and C is the feature dimension. Unlike macro-level action recognition tasks that typically employ a large number of memory slots, we specifically restrict K to a highly compact size for micro-gesture recognition. This crucial task-specific adaptation prevents the memory bank from overfitting to individual performer habits and high-frequency tracking jitters, ensuring that only robust, fine-grained motion semantics are distilled. Following the write-read memory operation in IB-Memory [9], the feature v is first written into the memory bank of its ground-truth class and then read out to obtain the purified representation z.

During memory writing, we first compute the similarity between v and each class-specific memory slot m_i^y:

$$s_i = \frac{v \cdot (m_i^y)^\top}{\|v\|_2 \|m_i^y\|_2}. \tag{1}$$

To encourage different memory slots to capture diverse intra-class variations, an inverse-weight writing strategy is adopted:

$$\beta_i = \frac{\exp(-s_i/\tau_w)}{\sum_{j=1}^{K} \exp(-s_j/\tau_w)}, \tag{2}$$

where τ_w is the writing temperature. The memory slot is then updated as:

$$m_i^y \leftarrow (1 - \eta\beta_i) m_i^y + \eta\beta_i v, \tag{3}$$

where η denotes the memory update rate. This inverse-weight update assigns larger writing weights to less activated slots, preventing the memory bank from repeatedly updating only a few dominant slots.

After memory writing, the updated memory bank is used for reading. The reading weight is computed by the temperature-scaled similarity:

$$\alpha_i = \frac{\exp(v \cdot (m_i^y)^\top/\tau_r)}{\sum_{j=1}^{K} \exp(v \cdot (m_j^y)^\top/\tau_r)}, \tag{4}$$

where τ_r is the reading temperature. The purified representation is obtained by weighted aggregation:

$$z = \sum_{i=1}^{K} \alpha_i m_i^y. \tag{5}$$

To enforce the compactness constraint of the information bottleneck, we regularize the latent space by minimizing the KL divergence $\mathcal{L}_{\text{compact}}$ between the posterior distribution of z and a standard normal prior $r(z)$.

The purified feature z is passed through an auxiliary classification head to obtain the semantic prediction distribution P_{IB}. Concurrently, the original backbone feature v is fed into the main classification head to produce P_{main}. To encourage the backbone to learn noise-invariant representations, we employ KL divergence to enforce consistency between P_{main} and P_{IB}:

$$\mathcal{L}_{\text{distill}} = D_{\text{KL}}(P_{\text{main}} \parallel P_{\text{IB}}). \tag{6}$$

Through this dual mechanism of variational compactness and distribution alignment, the backbone is encouraged to match the purified semantic distribution. After single-stream pre-training, the memory bank and auxiliary head are removed, and only the purified backbone weights are retained for subsequent joint-bone alignment and inference.

3.3 Joint-Bone Mutual Alignment

After IB-Memory pre-training, we denote the resulting purified joint and bone features as $\tilde{F}_J$ and $\tilde{F}_B$, respectively. To enhance the synergy between these heterogeneous skeleton representations, we introduce the Joint-Bone Mutual Alignment (JBMA) module.

First, the dual-stream features are mapped to probability distributions through their corresponding classification heads. A bi-directional KL alignment constraint is then applied to align the joint distribution P_J with the projected bone distribution $\hat{P}_B$, and vice versa:

$$\mathcal{L}_{\text{KL}} = \frac{1}{2}\left(D_{\text{KL}}(P_J \parallel \hat{P}_B) + D_{\text{KL}}(P_B \parallel \hat{P}_J)\right). \tag{7}$$

Meanwhile, the purified joint and bone features are concatenated along the channel dimension:

$$F_{\text{concat}} = [\tilde{F}_J, \tilde{F}_B]. \tag{8}$$

The concatenated representation is further processed by an independent fusion head and supervised with a cross-entropy loss $\mathcal{L}_{\text{concat}}$. During inference, the KL alignment branch is discarded, while the fusion head is retained to generate the skeleton branch prediction P_{sk}.

3.4 Stage-wise Optimization and Calibrated Fusion

The skeleton branch is optimized in two stages. In the first stage, the joint and bone streams are trained independently with the IB-Memory auxiliary branch:

$$\mathcal{L}_{\text{stage1}} = \mathcal{L}_{\text{main}} + \lambda_1\mathcal{L}_{\text{IB}} + \lambda_2\mathcal{L}_{\text{compact}} + \lambda_3\mathcal{L}_{\text{distill}}. \tag{9}$$

Here, $\mathcal{L}_{\text{main}}$ and $\mathcal{L}_{\text{IB}}$ denote the cross-entropy losses of the main head and IB auxiliary head, respectively.

In the second stage, the purified joint and bone streams are jointly optimized by JBMA:

$$\mathcal{L}_{\text{stage2}} = \mathcal{L}_{\text{concat}} + \lambda_4\mathcal{L}_{\text{KL}}. \tag{10}$$

For the final RGB-skeleton prediction, we use a lightweight calibrated fusion strategy. Given RGB logits z_{rgb} and skeleton logits z_{sk}, we first apply temperature scaling:

$$P_{\text{rgb}} = \text{softmax}(z_{\text{rgb}}/T_{\text{rgb}}), \quad P_{\text{sk}} = \text{softmax}(z_{\text{sk}}/T_{\text{sk}}). \tag{11}$$

A global modality fusion weight is selected on the validation set and fixed during testing:

$$P_{\text{final}} = \alpha P_{\text{rgb}} + (1 - \alpha) P_{\text{sk}}. \tag{12}$$

Here, α is a scalar coefficient controlling the contribution of the RGB prediction, while $1 - \alpha$ controls the contribution of the skeleton prediction. This calibrated fusion process introduces no additional network parameters and is used only as a final decision-level adjustment.

4 Experiments

4.1 Dataset

The iMiGUE [11,15] dataset comprises 18,499 labeled micro-gesture samples from 359 post-match press conference videos of professional tennis players, covering 31 micro-gesture categories and one non-micro-gesture category. The videos range from 0.5 to 26 minutes, with approximately 3.77 million frames in total. The dataset provides RGB videos and 2D skeleton joint data extracted via OpenPose. Following a cross-identity evaluation protocol, the dataset is split into 12,893 training, 777 validation, and 4,562 test samples.

4.2 Implementation Details

The framework is implemented in PyTorch and trained on two NVIDIA RTX 4090 GPUs.

- **Network Architecture and Initialization:** The RGB branch adopts a ViT-Base architecture with joint space-time attention. The tubelet size is set to $2 \times 16 \times 16$, and the weights are initialized with VideoMAE pretrained weights on Kinetics-400. The skeleton branch employs the SkateFormer architecture, with a feature embedding dimension of 96. During training, the IB-Memory auxiliary module assigns $K = 3$ class-specific memory slots to each category, resulting in 32×3 memory slots in total. Multiple slots within each category are used to model intra-class motion variations and performer-specific differences.
- **Data Preprocessing:** For the skeleton branch, the input consists of 24 upper-body joints, temporally sampled to 32 frames.
- **Optimizer and Hyperparameters:** We use the AdamW optimizer with a weight decay of 0.05 and a cosine annealing learning rate schedule. The initial learning rate for the skeleton branch is 5×10^{-4}, with a total batch size of 32. During model selection, we use the official training and validation split. For the final submission, after selecting the hyperparameters and checkpoints, we merge the training and validation samples to improve robustness.

4.3 Training Pipeline and Strategy

We adopt a three-step training strategy. First, the joint and bone streams are independently trained with IB-Memory for 25 epochs, including 5 warmup epochs. Second, IB-Memory is removed and JBMA is trained for 20 epochs, with the SkateFormer learning rate reduced to one-tenth of the initial value. Finally, the RGB and skeleton branches are frozen, and temperature-calibrated RGB-skeleton fusion is performed with $T_{\text{rgb}} = 1.3$ and $T_{\text{sk}} = 1.0$. Following the grid search strategy with a 0.01 step size on the validation set, the optimal global fusion weight α is determined to be 0.55.

4.4 Comparison with State-of-the-Arts

We compare the performance of our proposed framework with several state-of-the-art methods on the iMiGUE test set, including classic GCN-based skeleton models, spatio-temporal video models, and recent competitive solutions for micro-gesture recognition.

Table 1. Performance comparison on the iMiGUE test set in terms of Top-1 Accuracy.

Method	Modality	Top-1 Acc
ST-GCN [7]	Skeleton	46.38%
CTR-GCN [4]	Skeleton	53.02%
AAGCN [12]	Skeleton	54.73%
SkateFormer [5]	Skeleton	56.40%
Ours (Skeleton Branch)	**Skeleton**	**61.18%**
TSM [10]	RGB	58.77%
ViViT	RGB	67.84%
Ours (Late Fusion)	RGB + Skeleton	70.47%
Ours (Final Framework)	**RGB + Skeleton**	**71.37%**
MM-Gesture [6] (MiGA'25 1st)	Multi-modal	73.21%
GAIE-ViT [8] (MiGA'25 2nd)	RGB	67.70%

As shown in Table 1, our skeleton branch achieves 61.18% Top-1 accuracy, outperforming the SkateFormer baseline by 4.78 percentage points. With calibrated RGB-skeleton fusion, the final framework reaches 71.37%, demonstrating the effectiveness of skeleton purification, joint-bone alignment, and multimodal complementarity.

4.5 Ablation Study

To further investigate the contribution of the proposed skeleton optimization modules, we conduct ablation experiments on IB-Memory and JBMA. The final

RGB-skeleton fusion is treated as a lightweight decision-level calibration step and is not included as an independent ablation module.

Table 2. Ablation results of the proposed skeleton optimization modules.

Component	Configuration	Top-1 Acc	Gain
IB-Memory	Joint (Without IB-Memory)	57.23%	-
IB-Memory	**Joint (With IB-Memory)**	**59.73%**	**+2.50%**
IB-Memory	Bone (Without IB-Memory)	52.72%	-
IB-Memory	**Bone (With IB-Memory)**	**53.75%**	**+1.03%**
JBMA	Joint-Bone (Concat Fusion)	59.80%	-
JBMA	Joint-Bone (Late Fusion)	60.17%	-
JBMA	**JBMA (Concat + KL Alignment)**	**61.18%**	**+1.38%**

As shown in Table 2, IB-Memory consistently improves both individual skeleton streams. For the joint stream, IB-Memory improves the Top-1 accuracy from 57.23% to 59.73%, bringing a gain of 2.50%. For the bone stream, it improves the accuracy from 52.72% to 53.75%, yielding a gain of 1.03%. These results indicate that the memory-guided information bottleneck mechanism effectively suppresses task-irrelevant variations and provides more stable skeleton representations for fine-grained micro-gesture classification.

To address the contribution of the KL alignment term versus the feature concatenation, we isolate these components in our JBMA ablation. As shown in Table 2, simple feature concatenation supervised by a single fusion head (Joint-Bone Concat Fusion) achieves 59.80%, which yields negligible improvement over the Joint stream alone 59.73%. This indicates that without explicit semantic alignment, the network fails to effectively exploit the complementary bone information from the concatenated features. Late fusion of prediction logits performs slightly better at 60.17%. By introducing bi-directional KL consistency, our full JBMA achieves 61.18%, bringing a significant gain of 1.38% over plain concatenation and 1.01% over late fusion. This demonstrates that enforcing prediction-level semantic alignment is crucial for effective joint-bone feature fusion and unlocking their complementary potential.

5 Conclusion

This paper presents a skeleton-focused multimodal framework for micro-gesture recognition. IB-Memory suppresses task-irrelevant variations and enhances category-shared skeleton semantics, while JBMA improves joint-bone complementarity through bi-directional KL alignment and feature fusion. Experiments on iMiGUE show that our method achieves 71.37% Top-1 accuracy and ranks 3rd in the MiGA 2026 Challenge, demonstrating the effectiveness of skeleton feature purification and joint-bone mutual alignment.

Acknowledgments. The work is jointly supported by the National Natural Science Foundation of China under grants No.62472342, the Fundamental Research Funds for the Central Universities under grant No.QTZX25037.

Disclosure of Interests. The authors have no competing interests to declare that are relevant to the content of this article.

References

1. Alemi, A.A., Fischer, I., Dillon, J.V., Murphy, K.: Deep variational information bottleneck (2016). arXiv:1612.00410 arXiv preprint
2. Chen, G., et al.: Prototype learning for micro-gesture classification. arXiv preprint arXiv:2408.03097 (2024)
3. Chen, H., Shi, H., Liu, X., Li, X., Zhao, G.: SMG: a micro-gesture dataset towards spontaneous body gestures for emotional stress state analysis. Int. J. Comput. Vision **131**(6), 1346–1366 (2023)
4. Chen, Y., Zhang, Z., Yuan, C., Li, B., Deng, Y., Hu, W.: Channel-wise topology refinement graph convolution for skeleton-based action recognition. In: Proceedings of the IEEE/CVF International Conference on Computer Vision, pp. 13359–13368 (2021)
5. Do, J., Kim, M.: SkateFormer: skeletal-temporal transformer for human action recognition. In: European Conference on Computer Vision, pp. 401–420. Springer (2024)
6. Gu, J., Wang, F., Li, K., Wei, Y., Wu, Z., Guo, D.: MM-Gesture: towards precise micro-gesture recognition through multimodal fusion. arXiv preprint arXiv:2507.08344 (2025)
7. Han, H., et al.: STGCN: a spatial-temporal aware graph learning method for poi recommendation. In: 2020 IEEE International Conference on Data Mining (ICDM), pp. 1052–1057. IEEE (2020)
8. Hu, X., Pu, C., Li, Y., Xu, Y., Xie, K., Miao, Q.: Enhancing micro-gesture classification via global-aware importance estimation in vision transformer. In: MiGA@ IJCAI (2025)
9. Li, Y., Chen, H., Feng, G., Miao, Q.: Learning robust representations with information bottleneck and memory network for RGB-D-based gesture recognition. In: Proceedings of the IEEE/CVF International Conference on Computer Vision, pp. 20968–20978 (2023)
10. Lin, J., Gan, C., Han, S.: TSM: Temporal shift module for efficient video understanding. In: Proceedings of the IEEE/CVF International Conference on Computer Vision, pp. 7083–7093 (2019)
11. Liu, X., Shi, H., Chen, H., Yu, Z., Li, X., Zhao, G.: iMiGUE: an identity-free video dataset for micro-gesture understanding and emotion analysis. In: Proceedings of the IEEE/CVF Conference on Computer Vision and Pattern Recognition, pp. 10631–10642 (2021)
12. Shi, L., Zhang, Y., Cheng, J., Lu, H.: Skeleton-based action recognition with multi-stream adaptive graph convolutional networks. IEEE Trans. Image Process. **29**, 9532–9545 (2020)
13. Tishby, N., Pereira, F.C., Bialek, W.: The information bottleneck method (2000). physics/0004057 arXiv preprint

14. Wang, C.: The 4th MIGA-IJCAI challenge track 1. https://kaggle.com/competitions/the-4th-ei-mi-ga-ijcai-challenge-track-1 (2026). kaggle
15. Wang, C., Chen, H., Wei, H., Yang, Y., Chen, Y., Zhao, G.: iMiGUE-3K: a large-scale benchmark for micro-gesture analysis with self-supervised learning. arXiv preprint arXiv:2605.17179 (2026)
16. Weston, J., Chopra, S., Bordes, A.: Memory networks (2014). arXiv:1410.3916 arXiv preprint
17. Yan, S., Xiong, Y., Lin, D.: Spatial temporal graph convolutional networks for skeleton-based action recognition. In: Proceedings of the AAAI Conference on Artificial Intelligence, vol. 32 (2018)

A Multi-modal Framework with Cross-Subject Pseudo-labeling and Semantic Alignment for Micro-gesture Recognition

Haoran Zhang[1], Haokun Zhang[2], Pengyu Liu[1], Yujia Zhang[1], Weibao Xue[1], and Yanbin Hao[1](✉)

[1] School of Computer Science and Information Engineering, Hefei University of Technology (HFUT), Hefei, China
haoyanbin@hotmail.com
[2] School of Computer Science, University of Auckland(UOA), Auckland, New Zealand
hzha314@aucklanduni.ac.nz

Abstract. Micro-gestures (MGs) are spontaneous and subtle body movements that frequently convey hidden human emotions. Recognizing MGs in untrimmed videos remains highly challenging due to their extremely low signal-to-noise ratio, severe long-tailed class distribution, and the inherent domain shift encountered in cross-subject evaluation scenarios. In this paper, we propose a comprehensive multi-modal framework for Track 1 of the 4th MiGA-IJCAI Challenge. To capture fine-grained representations, we design a saliency-guided multi-modal extraction pipeline integrating 68-keypoint skeleton joint coordinates, 3D heatmap volumes, and high-resolution RGB visual features. We introduce a gentle square-root smoothed weighting mechanism paired with an Orthogonal Semantic Embedding Loss to protect tail classes without compromising overall recognition capabilities. More importantly, to bridge the cross-subject generalization gap, we propose a Cross-Modal Pseudo-Labeling (CMPL) strategy for unsupervised domain adaptation, which significantly boosts single-modal robustness. A temperature-scaled soft-voting mechanism is finally utilized to alleviate overconfidence during late fusion. Extensive experiments demonstrate that our framework achieves a competitive F1-score of 68.13%, securing the 4th place.

Keywords: Micro-gesture Recognition · Multi-modal Fusion · Unsupervised Domain Adaptation · Long-tailed Learning

1 Introduction

Micro-gestures (MGs) and micro-actions are spontaneous, subtle body movements (*e.g.*, Rubbing the nose) that reliably indicate hidden human emo-

H. Zhang and H. Zhang—Co-first author.

H. Chen et al. (Eds.): IJCAI 2026, CCIS 3120, pp. 46–59, 2027.
https://doi.org/10.1007/978-981-92-4529-1_5

tions [11]. Recognizing them automatically holds substantial practical significance for affective computing and clinical psychology [6,38,42,53]. However, Micro-gesture recognition (MGR) in untrimmed videos remains a formidable task, serving as the core objective of Track 1 in the 4th MiGA-IJCAI Challenge[1]. The primary difficulties of MGR stem from the intrinsically low action amplitudes and the minimal visual distinction between subtle movements and complex backgrounds, which collectively result in a severely low signal-to-noise ratio and make fine-grained recognition highly challenging.

To address these challenges, existing methodologies primarily rely on two parallel streams: skeleton-based and RGB-based approaches. Skeleton-based models [7,8,52] explicitly capture physical body topology but are sensitive to pose estimation errors and lack visual textures. On the other hand, RGB-based networks [1,4,43] encode rich spatio-temporal contexts but often struggle to isolate low-amplitude MGs from complex background noise. While multi-modal fusion attempts to integrate both [25], conventional paradigms often suffer from modality dominance. Although previous efforts have made improvements in mitigating the low signal-to-noise ratio of MGR, we observe that the extreme long-tailed distribution and cross-subject evaluation are two critical bottlenecks restricting the ultimate performance. Specifically, the severe long-tailed nature inherently biases the optimization process towards majority classes, resulting in poor discrimination of rare micro-gestures. Furthermore, the strict cross-subject protocol introduces significant domain shifts, leading to sharp performance degradation when the learned models are generalized to unseen individuals.

To overcome these bottlenecks, we propose a robust Multi-Modal Micro-gesture Framework that effectively integrates RGB and skeleton modalities to achieve noise suppression and feature complementarity. Specifically, rather than using raw full-frame videos, our vision branch performs saliency-guided cropping based on skeleton bounding boxes to isolate human subjects, thereby strictly filtering out background noise. These cropped high-resolution RGB clips are then processed by Video Swin Transformer [43] and R(2+1)D [50] to encode rich visual textures. In parallel, the skeleton branch leverages 68-keypoint sequences to capture physical topologies, utilizing a Decoupled Spatial-Temporal CNN for joint coordinates and a 3D-ResNet for rendered heatmaps [10]. To combat the extreme long-tailed distribution, we employ a square-root smoothed sampling strategy coupled with an Orthogonal Semantic Embedding Loss [33] to explicitly enlarge inter-class margins. Furthermore, to bridge the domain gap caused by unseen subjects, we design an iterative Cross-Modal Pseudo-Labeling (CMPL) [27,48]strategy, leveraging high-confidence consensus on the test set to formulate a super-dataset for unsupervised domain adaptation. Finally, multi-modal predictions are integrated through a temperature-scaled soft-voting fusion mechanism [17] to alleviate model overconfidence and guarantee optimal prediction reliability.

[1] The Kaggle competition page: https://www.kaggle.com/competitions/the-4th-ei-mi-ga-ijcai-challenge-track-1/leaderboard.

2 Related Work

2.1 Micro-gestures Dataset

Recognizing hidden emotions requires meticulously annotated and challenging benchmarks. The iMiGUE [42] dataset was specifically introduced as an identity-free dataset aimed at understanding suppressed emotions through spontaneous MGs. Unlike traditional gesture datasets that focus on illustrative macro-movements, iMiGUE captures involuntary gestures elicited during highly stressful scenarios (*e.g.*, post-match press conferences). It comprises 359 videos from 72 subjects across 28 countries, posing a significant challenge for cross-subject generalization.

To comprehensively analyze subtle human behaviors, the research community has progressively developed a series of specialized benchmarks. For instance, the MA-52 dataset [18] and related studies [15] were established to evaluate full-body spontaneous micro-actions during psychological interviews. Subsequently, the MMA-52 dataset [35] introduced multi-label annotations to support the complex task of concurrent micro-action detection in untrimmed videos. More recently, MA-Bench [30] was proposed to provide a finer-grained hierarchical evaluation platform. Driven by these rich datasets, the Micro-Action Analysis Challenges [19,34] have been hosted consecutively, successfully shifting the evaluation paradigms from isolated classification to continuous micro-gesture detection [13,39,40,47].

2.2 Multi-modal Action Recognition

Action recognition has significantly progressed with the evolution of deep learning architectures. For RGB modalities, 3D CNNs such as I3D [4], SlowFast [12], and R(2+1)D [50] laid the foundation for video modeling. Recently, Vision Transformers like VideoMAE [49], Video Swin Transformer [43], TimeSformer [1], and UniFormer [36] have shown superior capacity in capturing long-range visual dependencies. Furthermore, advanced contextual modeling and attention mechanisms [20,21] have been developed to efficiently capture spatial-temporal correlations in videos. Conversely, skeleton-based methods focus purely on the physical topology of human bodies, which has proven highly effective in general pose action recognition [32]. Graph Convolutional Networks (GCNs), such as ST-GCN [52] and CTR-GCN [7], Shift-GCN [8], and HD-GCN [28], have been widely adopted by recent micro-gesture recognition studies [5,16,31,33] to precisely extract structural motion features.

However, purely coordinate-based methods are frequently sensitive to topological noise and missing joints, which are prevalent in in-the-wild videos. To address this, Duan *et al.*proposed PoseC3D [10], rendering skeleton sequences into 3D pseudo-heatmaps. Meanwhile, to fully leverage heterogeneous features, recent works have also explored cross-modal feature enhancement [46] and self-supervised ensemble solutions [41] to boost micro-gesture recognition. Following the success of these mainstream paradigms, our framework explicitly leverages

Video Swin Transformer and R(2+1)D for high-resolution RGB visual textures, alongside PoseC3D and a Decoupled ST-CNN for robust structural joint cues, formulating a holistic multi-modal representation specifically tailored for fine-grained micro-gesture analysis.

2.3 Cross-Modal Pseudo-Labeling and Long-Tailed Learning

Real-world datasets inherently exhibit long-tailed class distributions. Traditional methods to calibrate such imbalance fall into re-sampling, re-weighting (*e.g.*, Focal Loss [37]), and decoupled representation learning [26,54]. While aggressive over-sampling is frequently utilized, we empirically observed it leads to severe mode collapse in extreme micro-gesture classification. Beyond standard calibration, recent works [33] utilize natural language word vectors to guide the latent feature space. Inspired by this, we integrate an Orthogonal Semantic Embedding Loss. Furthermore, cross-subject generalization requires aligning feature distributions across domains, often leveraging adversarial training [14] or source hypothesis transfer [29]. Pseudo-labeling [27,48] is a powerful technique to handle this. In our pipeline, we propose an iterative Cross-Modal Pseudo-Labeling (CMPL) strategy to significantly boost cross-subject generalization performance across diverse unseen domains.

3 Methodology

To effectively recognize micro-gestures, relying on a single modality is fundamentally insufficient due to the extremely low signal-to-noise ratio and subtle motion amplitudes. Our motivation is to construct a holistic representation that synergizes the explicit physical topology from skeletons, the spatio-temporal continuity from 3D heatmaps, and the fine-grained visual textures from RGB clips. Furthermore, to inherently address the extreme long-tailed class distribution and cross-subject domain shifts, we tightly integrate semantic regularization and unsupervised domain adaptation into our pipeline. The overall architecture of our proposed framework is explicitly illustrated in Fig. 1.

3.1 Saliency-Guided Multi-modal Encoding

Skeleton Joint Branch: Traditional 35-keypoint skeleton graphs fail to capture crucial facial micro-movements, such as biting lips or subtle jaw shifts. Therefore, we utilize the OpenPose toolkit [3] to extract $V = 68$ keypoints from the raw videos, deliberately incorporating 19 precise facial landmarks along with hand and body joints. Let $J \in \mathbb{R}^{C_{in} \times T \times V}$ denote the joint coordinates, where $C_{in} = 3$ represents $(x, y, \text{confidence})$ and T is the temporal length. To eliminate absolute spatial variations, we apply Center Normalization by subtracting the center joint coordinates before feeding them into the network. We process this sequence using a Decoupled Spatial-Temporal CNN, which independently applies (1×3) spatial convolutions and (5×1) temporal convolutions to efficiently capture the fine-grained spatial displacements of the continuous joint coordinate sequences with high structural fidelity.

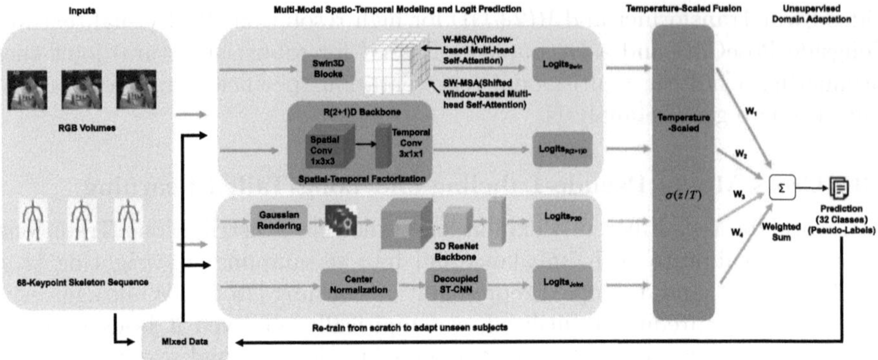

Fig. 1. The overall architecture of our proposed framework. First, raw RGB volumes and 68-keypoint skeletons are fed into four parallel branches (Swin3D, R(2+1)D, PoseC3D, and Decoupled ST-CNN) for multi-modal spatio-temporal modeling. Second, the generated logits are calibrated and aggregated via a Temperature-Scaled Fusion ($\sigma(z/T)$) to mitigate individual model overconfidence. Finally, an iterative pseudo-labeling loop forms a Data super-dataset for Unsupervised Domain Adaptation, enabling models to re-train from scratch and seamlessly adapt to unseen subjects

3D Heatmap Branch (PoseC3D): Purely coordinate-based models are intrinsically vulnerable to coordinate jittering and tracking failures, which can easily overwhelm the low-amplitude motion signals of micro-gestures. To construct a more resilient structural representation, we adopt PoseC3D [10]. The 68 joint coordinates are rendered into a 3D heatmap volume $\mathcal{H} \in \mathbb{R}^{V \times T \times H \times W}$, where $V = 68$ represents independent channels for each keypoint, and H, W represent the spatial resolution. A 3D-ResNet50 is adopted as the backbone. By interpreting coordinates as Gaussian heatmaps, 3D convolutions effortlessly bridge short-term joint disappearances through spatial-temporal continuity. We also enable horizontal flipping during target generation to effectively augment the spatial receptive field during the network training phase.

Saliency-Guided Vision Branch: Skeletons fundamentally discard appearance context, making it impossible to capture subtle facial deformations (*e.g.*, biting lips) or fine-grained hand-body interactions (*e.g.*, adjusting a collar). To complement this structural deficiency, the RGB modality is indispensable for providing dense visual cues. However, feeding entire raw frames introduces severe background noise. Thus, we introduce a Saliency-Guided Cropping module. We dynamically crop the Region of Interest (ROI) bounding boxes guided by the extreme boundaries of the skeleton coordinates (with a 20% safety margin). These cropped subjects are magnified to 224×224 high resolution. The visual features are then extracted using Video Swin Transformer (Swin3D) [43] and R(2+1)D [50] pre-trained on Kinetics-400 to extract rich spatial-temporal visual textures from the cropped regions.

3.2 Rethinking Long-Tailed Calibration in Extreme Imbalance

The iMiGUE dataset inherently exhibits a severe long-tailed class distribution. As explicitly illustrated in Fig. 2, a few head classes (particularly the non-micro-gesture background category) naturally dominate the dataset, while the vast majority of target micro-gesture categories suffer from extremely scarce training instances. To address this severe data imbalance issue, we propose a soft regularization approach:

1) Square-Root Smoothed Weighting: To prevent gradient explosion for rare classes—a common issue when directly applying inverse-frequency weighting [9]—we calculate the occurrence frequency N_c for each class c. We define a smoothed penalty weight $w_c = 1/\sqrt{N_c + 1}$. This weight is effectively integrated into the Focal Loss [37], gently encouraging the model to attend to minority classes without memorizing specific instances or overfitting to background noise, which is a crucial principle for robust long-tailed learning [2].

2) Orthogonal Semantic Embedding Loss: Due to the high visual similarity among micro-gestures, we introduce an auxiliary semantic loss, inspired by recent advancements in visual-semantic alignment [33]. Let $f \in \mathbb{R}^d$ be the extracted visual feature before the classifier. We define a non-trainable semantic matrix $M \in \mathbb{R}^{K \times d}$ representing $K = 32$ action classes with orthogonal initialization (generated via QR decomposition of a random matrix). While fixed anchors might impose strict constraints, they act as a powerful structural prior (analogous to prototype learning [5,31]) to prevent the model from overfitting to the background noise of tail classes. As shown in Fig. 3, the semantic alignment loss is formulated as the Mean Squared Error (MSE) between the feature and the target semantic anchor:

$$\mathcal{L}_{sem} = \text{MSE}(f, M_y) = \frac{1}{d}\sum_{i=1}^{d}(f_i - M_{y,i})^2 \tag{1}$$

where y is the ground-truth label. This explicit constraint forces the network to enlarge the inter-class distance among highly confusing micro-gestures within the latent semantic embedding space [33].

3.3 Cross-Modal Pseudo-Labeling (CMPL)

To bridge the domain gap inherent in the cross-subject testing protocol, we introduce a powerful UDA strategy. We first aggregate our baseline multi-modal models to generate high-confidence predictions $\hat{Y}_{test}$ on the unlabeled test set $\mathcal{D}_{test}$. Specifically, we select target samples whose maximum softmax probability across all 32 action categories exceeds a stringent confidence threshold α. Based on empirical evaluations, we set $\alpha = 0.65$ to guarantee the reliability of the generated annotations. We then treat these filtered predictions as pseudo-labels

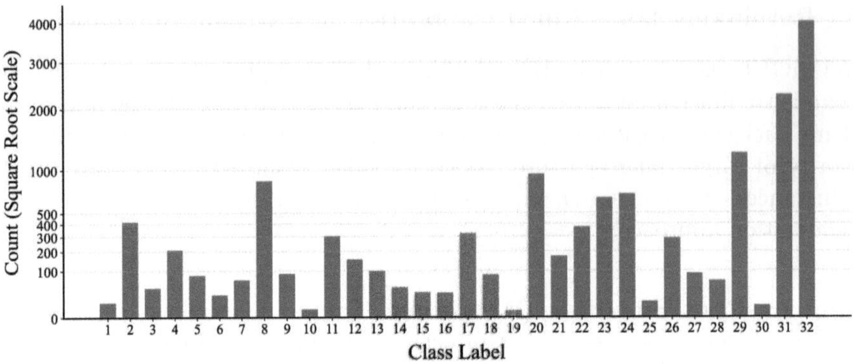

Fig. 2. Illustration of the extreme long-tailed class distribution in the iMiGUE dataset. The severe imbalance ratio (> 2000 : 1) often leads to catastrophic mode collapse if traditional aggressive re-sampling techniques are applied.

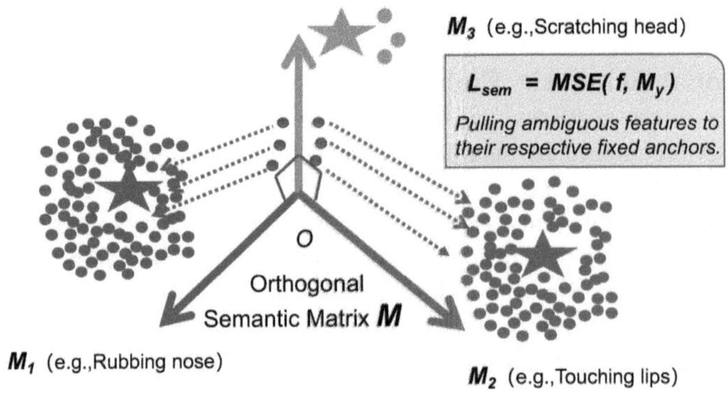

Fig. 3. A conceptual diagram illustrating the Orthogonal Semantic Embedding Loss. This mechanism explicitly pulls ambiguous visual features toward their respective fixed orthogonal anchors, effectively preventing minority tail classes from being swallowed by dominant majority head classes within the latent feature space.

and inject them into the training set, forming an augmented training set $\mathcal{D}_{aug} = \mathcal{D}_{train} \cup \mathcal{D}_{val} \cup \mathcal{D}_{test}(\hat{Y}_{test})$.

By performing a single round of retraining from scratch on this super-dataset, the models implicitly learn the idiosyncratic behavioral nuances and fine-grained micro-action patterns of the unseen subjects while avoiding confirmation bias. This exposure drastically shifts the data distribution and lifts the single-modal accuracy ceiling without requiring any manual annotations or target domain labels.

3.4 Temperature-Scaled Multi-modal Fusion

Late fusion through direct probability addition often suffers from the "veto effect", where an overconfident but incorrect model dominates the ensemble. To address this, we apply Temperature Scaling before the softmax operation to calibrate the logits L_m from model m:

$$P_m^{(c)} = \frac{\exp(L_m^{(c)}/\tau)}{\sum_{j=1}^{K} \exp(L_m^{(j)}/\tau)}, \tag{2}$$

where τ is the temperature parameter (set to 2.5 via grid search on the validation set). This technique softens sharp probability distributions, democratizing the voting process.

The final fused probability is obtained via a weighted sum:

$$P_{final}^{(c)} = \sum_{m=1}^{M} W_m P_m^{(c)}, \tag{3}$$

where the modality weights W_m are strictly determined through grid search based on validation performance, allowing fine-grained corrective cues from auxiliary models to influence the final prediction effectively and seamlessly mitigate the veto effect.

4 Experiments

4.1 Dataset and Evaluation Metric

The iMiGUE dataset [42] is collected to evaluate hidden emotion understanding through micro-gestures. It comprises 32 classes of spontaneous micro-gestures. We utilize a strict cross-subject evaluation protocol, dividing the involved subjects into distinct training, validation, and testing groups without any overlapping identities. Furthermore, we employ the Top-1 Accuracy as our primary evaluation metric, which strictly corresponds to the Micro F1-Score within this multi-class setting.

4.2 Implementation Details

The **Skeleton Joint model** processes center-normalized 68-keypoint coordinate sequences, which are uniformly sampled to 64 frames. To optimize this network, we utilize the AdamW optimizer [45] paired with a Cosine Annealing learning rate schedule [44]. For the **PoseC3D model** [10], 3D pseudo-heatmaps are rendered at a spatial resolution of 56×56 across 48 temporal frames. This branch is optimized using Stochastic Gradient Descent (SGD) for 100 epochs, employing a large batch size of 64 per GPU to ensure stable gradient convergence. Regarding the **RGB Vision models**, each cropped video clip is temporally sampled to yield 24 to 32 frames at a spatial resolution of 224×224. To further bolster the robustness of visual representations against spatial variances, we perform Test-Time Augmentation (TTA) during the inference phase by averaging the output logits from the original and horizontally flipped video sequences.

4.3 Main Results

Table 1 illustrates the performance comparison with top-performing methods from current and previous challenges on the iMiGUE cross-subject test set. Benefiting from the robust multi-modal feature encoding and the powerful cross-subject pseudo-labeling adaptation, our proposed ultimate ensemble achieves a highly competitive accuracy of 68.13%, securing the 4th place in the challenge.

Table 1. Top submissions of the past four MiGA Track-1 editions (2023–2026) on the iMiGUE test split, ranked by top-1 accuracy. Our submission (AIM) is highlighted in bold. Modalities: **J** = joint, **L** = limb, **R** = RGB, **T** = Taylor video, **F** = optical flow, **D** = depth video, **H** = heatmap.

Rank	Team	Core Methodology	Modality	Acc. (%)
MiGA'23 1st	HFUT-VUT [33]	PoseConv3D	**J+L**	64.12
MiGA'23 2nd	NPU-Stanford [23]	Hyperformer	**J**	63.02
MiGA'24 1st	HFUT-VUT [5]	PoseConv3D	**J+L+R**	70.25
MiGA'24 2nd	NPU-MUCIS [24]	Res2Net3D+GCN	**J+R**	70.19
MiGA'25 1st	HFUT-VUT [16]	PoseConv3D+VideoSwinT	**J+L+R+ T+F+D**	73.21
MiGA'25 2nd	awuniverse [22]	ViT-Base+GAIE	**R**	68.70
MiGA'25 3rd	Lonelysheep [51]	PoseConv3D	**J+L**	67.01
MiGA'26 1st	XInsight Lab [41]	PoseConv3D+VideoSwinT +SMILE	**J+L+R+D**	74.41
MiGA'26 2nd	Team YUV	-	-	74.24
MiGA'26 3rd	Team xd	-	-	71.37
MiGA'26 4th	**AIM (Ours)**	**Decoupled ST-CNN + PoseC3D + Swin3D & R(2+1)D**	**J+H+R**	**68.13**

* *Note:* Our proposed framework integrates four complementary branches: pseudo-labeled Decoupled ST-CNN (Skeleton Joint), PoseC3D (3D Heatmaps), and saliency-cropped Swin3D & R(2+1)D (RGB), fused via temperature-scaled soft voting.

4.4 Ablation Study

In Table 2, we report the results of our ablation experiments conducted on several baselines to demonstrate the overall effectiveness of our proposed framework. Furthermore, we illustrate the specific effectiveness of the proposed spatial-temporal encoding, semantic alignment, and cross-modal pseudo-labeling by incrementally integrating these modules.

Table 2. Ablation study of our proposed framework on the iMiGUE Test Set. Modalities: **J** = Joint, **H** = Pseudo-Heatmap, **R** = RGB.

Method / Component	Modality	Accuracy (%)
Baseline ST-GCN (35-keypoint)	**J**	48.00
Baseline RGB R(2+1)D (without cropping)	**R**	46.90
High-Resolution Cropped RGB R(2+1)D (w/ TTA)	**R**	58.89
PoseC3D (with 68 facial & body keypoints)	**H**	60.58
Skeleton Joint (Decoupled ST-CNN + 68kp + Sem-Loss)	**J**	61.70
Skeleton Joint (with Cross-Modal Pseudo-Labeling)	**J**	**66.65**
Ours	**J + H + R**	**68.13**

Effectiveness of Spatial-Temporal Encoding and Semantic Alignment: We evaluate the combined effect of upgrading the architecture to Decoupled ST-CNN, increasing keypoints from 35 to 68 (introducing 19 facial landmarks), and applying the Semantic Alignment Loss. As shown in Table 2, this joint optimization boosted the single-modal skeleton accuracy from 48.00% to 61.70%, demonstrating the effectiveness of incorporating dense facial cues and semantic regularization for fine-grained gesture recognition.

Impact of Cross-Modal Pseudo-Labeling: The application of cross-modal pseudo-labeling further enhanced the model's performance. By retraining the network from scratch on the combined dataset equipped with pseudo-labeled test samples, the accuracy of the single-modal Skeleton Joint model increased to 66.65%. This improvement indicates that implicitly aligning the target domain distribution via pseudo-labels effectively mitigates the domain shift issue inherent in cross-subject evaluations.

Comparison of Late Fusion Strategies: We evaluated different late fusion strategies and empirically found that Stacking Classifiers (*e.g.*, LightGBM) are prone to overfitting the limited validation data (777 samples). This introduces extra learnable parameters that degrade generalization on the unseen test set, yielding an accuracy of approximately 56%. In contrast, the Temperature-Scaled Soft Voting method explicitly calibrates the prediction confidence without introducing additional trainable parameters. Consequently, it exhibits better stability under cross-subject domain shifts and achieves the optimal performance in our final multi-modal ensemble.

5 Conclusion

In this paper, we presented our solution for the Micro-gesture Classification track of the 4th MiGA-IJCAI Challenge. Our approach is based on a robust multimodal framework, with the introduction of an iterative cross-modal pseudo-labeling strategy to alleviate the severe domain shift commonly observed in cross-subject evaluation. Additionally, we enhance the model's ability to capture fine-grained human behaviors by integrating 68-keypoint skeleton joint features, 3D pseudo-heatmaps, and high-resolution RGB visual contexts, while addressing the extreme long-tailed distribution via an Orthogonal Semantic Alignment Loss. Our final ensemble model achieved a Top-1 Accuracy of 68.13% on the test set of the iMiGUE dataset. As a result, our method ranked 4th in this challenge track.

Acknowledgments. This work is supported by the National Natural Science Foundation of China (No. 62472393).

References

1. Bertasius, G., Wang, H., Torresani, L.: Is space-time attention all you need for video understanding? In: International Conference on Machine Learning, pp. 813–824. PMLR (2021)
2. Cao, K., Wei, C., Gaidon, A., Arechiga, N., Ma, T.: Learning imbalanced datasets with label-distribution-aware margin loss. In: Advances in Neural Information Processing Systems, vol. 32 (2019)
3. Cao, Z., Hidalgo, G., Simon, T., Wei, S.E., Sheikh, Y.: OpenPose: realtime multi-person 2D pose estimation using part affinity fields. IEEE Trans. Pattern Anal. Mach. Intell. **43**(1), 172–186 (2019)
4. Carreira, J., Zisserman, A.: Quo vadis, action recognition? A new model and the Kinetics dataset. In: Proceedings of the IEEE Conference on Computer Vision and Pattern Recognition, pp. 6299–6308 (2017)
5. Chen, G., et al.: Prototype learning for micro-gesture classification. arXiv preprint arXiv:2408.03097 (2024)
6. Chen, H., Shi, H., Liu, X., Li, X., Zhao, G.: SMG: a micro-gesture dataset towards spontaneous body gestures for emotional stress state analysis. Int. J. Comput. Vision **131**(5), 1346–1366 (2023)
7. Chen, Y., Zhang, Z., Yuan, C., Li, B., Deng, Y., Hu, W.: Channel-wise topology refinement graph convolution for skeleton-based action recognition. In: Proceedings of the IEEE/CVF International Conference on Computer Vision, pp. 13359–13368 (2021)
8. Cheng, K., Zhang, Y., He, X., Chen, W., Cheng, J., Lu, H.: Skeleton-based action recognition with shift graph convolutional network. In: Proceedings of the IEEE/CVF Conference on Computer Vision and Pattern Recognition, pp. 183–192 (2020)
9. Cui, Y., Jia, M., Lin, T.Y., Song, Y., Belongie, S.: Class-balanced loss based on effective number of samples. In: Proceedings of the IEEE/CVF Conference on Computer Vision and Pattern Recognition, pp. 9268–9277 (2019)
10. Duan, H., Zhao, Y., Chen, K., Lin, D., Dai, B.: Revisiting skeleton-based action recognition. In: Proceedings of the IEEE/CVF Conference on Computer Vision and Pattern Recognition, pp. 2969–2978 (2022)
11. Ekman, P., Friesen, W.V.: Nonverbal leakage and clues to deception. Psychiatry **32**(1), 88–106 (1969)

12. Feichtenhofer, C., Fan, H., Malik, J., He, K.: SlowFast networks for video recognition. In: Proceedings of the IEEE/CVF International Conference on Computer Vision, pp. 6202–6211 (2019)
13. Filntisis, P.P., Efthymiou, N., Potamianos, G., Maragos, P.: Emotion understanding in videos through body, context, and visual-semantic embedding loss. In: Bartoli, A., Fusiello, A. (eds.) ECCV 2020. LNCS, vol. 12535, pp. 747–755. Springer, Cham (2020). https://doi.org/10.1007/978-3-030-66415-2_52
14. Ganin, Y., Ustinova, E., Ajakan, H., Germain, P., Larochelle, H., Laviolette, F., Marchand, M., Lempitsky, V.: Domain-adversarial training of neural networks. J. Mach. Learn. Res. **17**(1), 2030–2096 (2016)
15. Gu, J., et al.: Motion matters: motion-guided modulation network for skeleton-based micro-action recognition. In: Proceedings of the 33rd ACM International Conference on Multimedia, pp. 5461–5470 (2025)
16. Gu, J., Wang, F., Li, K., Wei, Y., Wu, Z., Guo, D.: MM-gesture: towards precise micro-gesture recognition through multimodal fusion. arXiv preprint arXiv:2507.08344 (2025)
17. Guo, C., Pleiss, G., Sun, Y., Weinberger, K.Q.: On calibration of modern neural networks. In: International Conference on Machine Learning, pp. 1321–1330. PMLR (2017)
18. Guo, D., Li, K., Hu, B., Zhang, Y., Wang, M.: Benchmarking micro-action recognition: dataset, methods, and applications. IEEE Trans. Circuits Syst. Video Technol. **34**(7), 6238–6252 (2024)
19. Guo, D., et al.: MAC 2024: micro-action analysis grand challenge. In: Proceedings of the 32nd ACM International Conference on Multimedia, pp. 11304–11305 (2024)
20. Hao, Y., et al.: Attention in attention: modeling context correlation for efficient video classification. IEEE Trans. Circuits Syst. Video Technol. **32**(10), 7120–7132 (2022)
21. Hao, Y., Zhang, H., Ngo, C.W., He, X.: Group contextualization for video recognition. In: Proceedings of the IEEE/CVF Conference on Computer Vision and Pattern Recognition, pp. 928–938 (2022)
22. Hu, X., Pu, C., Li, Y., Xie, K., Qiguang, M.: Enhancing micro-gesture classification via global-aware importance estimation in vision transformer. In: Proceedings of the MiGA Workshop at IJCAI (2025)
23. Huang, H., Guo, X., Peng, W., Xia, Z.: Micro-gesture classification based on ensemble hypergraph convolution transformer. In: Proceedings of the MiGA Workshop at IJCAI (2023)
24. Huang, H., Wang, Y., Kerui, L., Xia, Z.: Multi-modal micro-gesture classification via multiscale heterogeneous ensemble network. In: Proceedings of the MiGA Workshop at IJCAI (2024)
25. Joze, H.R.V., Shaban, A., Iuzzolino, M.L., Koishida, K.: MMTM: multimodal transfer module for CNN fusion. In: Proceedings of the IEEE/CVF Conference on Computer Vision and Pattern Recognition, pp. 13289–13299 (2020)
26. Kang, B., et al.: Decoupling representation and classifier for long-tailed recognition. In: International Conference on Learning Representations (2020)
27. Lee, D.H.: Pseudo-label: the simple and efficient semi-supervised learning method for deep neural networks. In: Workshop on Challenges in Representation Learning, ICML, vol. 3, p. 896 (2013)
28. Lee, J., Lee, M., Lee, D., Lee, S.: Hierarchically decoupled graph convolutional network for skeleton-based action recognition. In: Proceedings of the IEEE/CVF International Conference on Computer Vision, pp. 10444–10453 (2023)

29. Li, J., Qiu, R., Chen, S.: Do we really need to access the source data? Source hypothesis transfer for unsupervised domain adaptation. In: International Conference on Machine Learning, pp. 6028–6039. PMLR (2020)
30. Li, K., Gu, J., Wang, F., Wu, Z., Fan, H., Guo, D.: MA-Bench: towards fine-grained micro-action understanding. arXiv preprint arXiv:2603.26586 (2026)
31. Li, K., et al.: Prototypical calibrating ambiguous samples for micro-action recognition. In: Proceedings of the AAAI Conference on Artificial Intelligence, vol. 39, pp. 4815–4823 (2025)
32. Li, K., Guo, D., Chen, G., Liu, F., Wang, M.: Data augmentation for human behavior analysis in multi-person conversations. In: Proceedings of the 31st ACM International Conference on Multimedia, pp. 9516–9520 (2023)
33. Li, K., Guo, D., Chen, G., Peng, X., Wang, M.: Joint skeletal and semantic embedding loss for micro-gesture classification. arXiv preprint arXiv:2307.10624 (2023)
34. Li, K., et al.: MAC 2025: the 2nd micro-action analysis grand challenge. In: Proceedings of the 33rd ACM International Conference on Multimedia, pp. 14216–14221 (2025)
35. Li, K., et al.: MMAD: multi-label micro-action detection in videos. In: Proceedings of the IEEE/CVF International Conference on Computer Vision, pp. 13225–13236 (2025)
36. Li, K., et al.: UniFormer: unifying convolution and self-attention for visual recognition. In: International Conference on Learning Representations (2022)
37. Lin, T.Y., Goyal, P., Girshick, R., He, K., Dollár, P.: Focal loss for dense object detection. In: Proceedings of the IEEE International Conference on Computer Vision, pp. 2980–2988 (2017)
38. Liu, P., et al.: A survey on fMRI-based brain decoding for reconstructing multimodal stimuli. arXiv preprint arXiv:2503.15978 (2025)
39. Liu, P., Li, K., Wang, F., Wei, Y., She, J., Guo, D.: Online micro-gesture recognition using data augmentation and spatial-temporal attention. arXiv preprint arXiv:2507.09512 (2025)
40. Liu, P., et al.: Micro-gesture online recognition using learnable query points. arXiv preprint arXiv:2407.04490 (2024)
41. Liu, T., et al.: Self-supervised learning matters: a simple ensemble solution for micro-gesture recognition. arXiv preprint arXiv:2606.09261 (2026)
42. Liu, X., Shi, H., Chen, H., Yu, Z., Li, X., Zhao, G.: iMiGUE: an identity-free video dataset for micro-gesture understanding and emotion analysis. In: Proceedings of the IEEE/CVF Conference on Computer Vision and Pattern Recognition, pp. 10631–10642 (2021)
43. Liu, Z., et al.: Video swin transformer. In: Proceedings of the IEEE/CVF Conference on Computer Vision and Pattern Recognition, pp. 3202–3211 (2022)
44. Loshchilov, I., Hutter, F.: SGDR: stochastic gradient descent with warm restarts. In: International Conference on Learning Representations (2017)
45. Loshchilov, I., Hutter, F.: Decoupled weight decay regularization. In: International Conference on Learning Representations (2019)
46. Shang, T., Hao, Y., Pei, M., Li, K., Ben, H., Wang, S.: Cross-modal feature enhancement and contrastive alignment for micro-gesture recognition. In: Kittler, J., et al. (eds.) Chinese Conference on Pattern Recognition and Computer Vision (PRCV), pp. 203–217. Springer, Singapore (2025). https://doi.org/10.1007/978-981-95-5676-2_14
47. Shen, X., Li, K., Wang, F., Qian, W., Jiang, J., Guo, D.: Spatial-temporal decoupled adapter for micro-gesture online recognition. arXiv preprint arXiv:2606.07355 (2026)

48. Sohn, K., et al.: FixMatch: simplifying semi-supervised learning with consistency and confidence. In: Advances in Neural Information Processing Systems, vol. 33, pp. 596–608 (2020)
49. Tong, Z., Song, Y., Wang, J., Wang, L.: VideoMAE: masked autoencoders are data-efficient learners for self-supervised video pre-training. In: Advances in Neural Information Processing Systems, vol. 35, pp. 10078–10093 (2022)
50. Tran, D., Wang, H., Torresani, L., Ray, J., LeCun, Y., Paluri, M.: A closer look at spatiotemporal convolutions for action recognition. In: Proceedings of the IEEE Conference on Computer Vision and Pattern Recognition, pp. 6450–6459 (2018)
51. Xu, H., Cheng, L., Wang, Y., Tang, S., Zhong, Z.: Towards fine-grained emotion understanding via skeleton-based micro-gesture recognition. In: Proceedings of the MiGA Workshop at IJCAI (2025)
52. Yan, S., Xiong, Y., Lin, D.: Spatial temporal graph convolutional networks for skeleton-based action recognition. In: Proceedings of the AAAI Conference on Artificial Intelligence, vol. 32 (2018)
53. Yan, W.J., et al.: CASME II: a database for spontaneous macro-expression and micro-expression spotting and recognition. PLoS ONE **9**(1), e86041 (2014)
54. Zhou, B., Cui, Q., Wei, X.S., Chen, Z.M.: BBN: bilateral-branch network with cumulative learning for long-tailed visual recognition. In: Proceedings of the IEEE/CVF Conference on Computer Vision and Pattern Recognition, pp. 9719–9728 (2020)

Spatial-Temporal Decoupled Adapter for Micro-gesture Online Recognition

Xucheng Shen[1], Kun Li[2(✉)], Fei Wang[1,3,4], Wei Qian[1], Jin Jiang[2], and Dan Guo[1,3]

[1] Hefei University of Technology, Hefei, China
[2] United Arab Emirates University, Al Ain, United Arab Emirates
kunli.hfut@gmail.com
[3] Institute of Artificial Intelligence, Hefei Comprehensive National Science Center, Hefei, China
[4] Anhui Evolution Technology Co., Ltd., Hefei, China

Abstract. Micro-gesture online recognition aims to temporally localize and classify subtle gestures in untrimmed videos. Owing to their extremely short duration, low motion amplitude, and ambiguous visual cues, capturing discriminative spatiotemporal representations remains highly challenging. Existing parameter-efficient adapters typically employ a single branch to model spatial and temporal cues jointly, which may fail to capture the fine-grained patterns of micro-gestures. To address this limitation, we propose a Spatial-Temporal Decoupled Adapter that decomposes video adaptation into independent temporal and spatial branches via lightweight depthwise convolutions. In addition, to alleviate the long-tailed class distribution inherent in the benchmark dataset, we introduce an Adaptive Soft Balanced Augmentation method, which dynamically allocates augmentation intensity based on class rarity and learning difficulty, without manual thresholds. Our method achieves an F1 score of 0.43808, ranking 1st in Track 2 of the 4th EI-MiGA-IJCAI Challenge.

Keywords: Micro-Gesture Recognition · Data Augmentation · Parameter-Efficient Fine-Tuning

1 Introduction

Micro-gestures [6,13,34] are subtle and frequently unconscious body movements, primarily involving the hands, fingers, and arms, that arise spontaneously in the course of interpersonal communication. These inconspicuous behaviors provide valuable insights into human psychological states and emotional conditions, making them increasingly important for affective computing [9,39,40,47], psychological assessment [4,37,38], and intelligent human-computer interaction [28].

As a highly challenging task, micro-gesture online recognition requires both temporal localization and category classification of micro-gesture instances in

H. Chen et al. (Eds.): IJCAI 2026, CCIS 3120, pp. 60–76, 2027.
https://doi.org/10.1007/978-981-92-4529-1_6

untrimmed videos. Compared to traditional action recognition [15,16] or temporal action detection [26,52], this task places greater emphasis on distinguishing between fine-grained micro-gesture categories and precisely identifying the start and end times of each instance. The core difficulty lies in capturing subtle differences among micro-gestures to distinguish fine-grained categories and accurately determine the start and end timestamps of action segments.

Traditional temporal action detection methods [26,32,54] typically rely on a two-stage paradigm, where video features are extracted offline and subsequently used to train a detector. Such a design prevents the backbone from being optimized by the detection loss, potentially limiting representation learning. End-to-end training overcomes this by jointly learning the backbone and detection head, which theoretically yields stronger representations. However, fully fine-tuning a large model like VideoMAE-g requires extremely high GPU memory that common hardware cannot afford.

Recent advances [1,31] have explored parameter-efficient fine-tuning of pretrained vision encoders for temporal action detection, achieving an effective balance between performance and computational efficiency. A common strategy is to insert lightweight adapter modules into frozen backbone layers, enabling the pretrained model to efficiently adapt to the downstream detection task with minimal additional parameters. Most existing designs adopt a single-branch structure that processes spatial and temporal information within a unified pathway, which may limit the model's ability to independently capture local spatial patterns and long-range temporal dynamics. To address this issue, we propose a Spatial-Temporal Decoupled Adapter that decouples spatial and temporal modeling into parallel streams, allowing each branch to specialize in its respective dimension before merging their complementary representations.

In addition to the challenge of modeling subtle spatio-temporal cues, micro-gesture recognition also suffers from severe data imbalance. As the official benchmark of the EI-MiGA-IJCAI Challenge Track 2, the SMG dataset [6] consists of 40 untrimmed videos spanning 16 micro-gesture categories and one non-micro-gesture category, with a highly skewed class distribution where several categories contain significantly fewer samples than the majority classes. To alleviate this long-tail issue, we introduce an Adaptive Soft Balanced Augmentation method, which dynamically adjusts augmentation intensity for each class based on effective sample counts and learning difficulty, without relying on manual thresholds.

In summary, the contributions of our framework are as follows:

- We propose a Spatial-Temporal Decoupled Adapter for parameter-efficient fine-tuning, enabling enhanced fine-grained spatio-temporal representation learning for micro-gesture recognition.
- We introduce an Adaptive Soft Balanced Augmentation method to improve robustness against long-tail class imbalance by adaptively determining augmentation intensity according to class rarity and learning difficulty.
- Experiments on the SMG dataset demonstrate the effectiveness of both components for micro-gesture online recognition.

2 Related Work

2.1 Micro-gesture Analysis

Micro-gesture analysis [6,34], including micro-gesture recognition [3,11,22,41], micro-gesture online detection [29,30], and behavior-based emotion understanding [49], has attracted increasing research attention with the development of dedicated datasets [12,20,24,45], advanced methods [10,21,33], and competitions [5,13,23]. The iMiGUE [34] is the first public dataset for micro-gesture understanding and emotion analysis, containing over 18,000 video clips across 32 categories. SMG [6] focuses on spontaneous micro-gestures under psychological stress with 16 MG categories and one non-MG category. MA-52 [12] extends to 52 micro-action categories, covering fine-grained natural actions in daily communication. MMA-52 [24] introduces multi-label annotations to better reflect action co-occurrence in real-world scenarios. These datasets establish standardized benchmarks for research on fine-grained non-verbal behavior.

Micro-gesture online recognition is challenging due to subtle motion patterns and sparse temporal distribution. Early methods adopted skeleton-based approaches: Guo *et al.* [14] combined GCNs with multi-scale Transformers, leveraging the structural prior of human joints to capture fine-grained spatial dependencies, though skeleton extraction errors in occluded scenarios limited robustness. Wang *et al.* [48] constructed a dual-stream architecture processing RGB and skeleton modalities in parallel, fusing appearance and structural cues through cross-modal attention, which demonstrated that complementary modalities can compensate for the ambiguity inherent in single-source inputs.

Recent methods have shifted toward large-scale pretrained video backbones, exploiting their rich generalizable representations to reduce reliance on handcrafted features. Liu *et al.* [30] introduced learnable query points coupled with Mamba blocks, replacing conventional temporal modeling with selective state-space scanning that offers linear complexity while preserving long-range context, achieving competitive results using only RGB input. In MiGA 2025, Liu *et al.* [29] proposed data augmentation strategies and spatial-temporal attention modules built upon DyFADet with VideoMAEv2-g features. Their augmentation pipeline alleviated the severe class imbalance in micro-gesture datasets, while the attention modules sharpened boundary localization, achieving an F1 score of 0.38027. Meng *et al.* [35] adopted a two-stage pipeline using a frozen VideoMAE-g encoder and DyFADet, validating that combining pretrained visual representations with dynamic sequence modeling advances micro-gesture understanding in complex real-world settings.

2.2 Parameter-Efficient Fine-Tuning for Video Understanding

Parameter-efficient fine-tuning has become a popular strategy for adapting large pretrained models to downstream tasks with limited trainable parameters. Existing PEFT methods mainly include prompt tuning [19,25], low-rank adaptation [18], and adapter-based tuning [7,17]. Among these approaches, adapter tuning is a widely used PEFT strategy that freezes most parameters of the pretrained backbone and inserts small learnable modules into intermediate layers. During fine-tuning, adapters allow pretrained vision representations to be adapted to downstream tasks with minimal additional parameters while preserving the general knowledge learned during large-scale pretraining, and learning task-specific transformations while keeping most pretrained parameters fixed.

Recent advances [7,36,51] have demonstrated the strong potential of adapter-based PEFT for adapting large pretrained vision models to downstream video understanding tasks. AdaptFormer [7] introduces lightweight adapter modules into Vision Transformers, enabling efficient task adaptation while preserving the generalization capability of pretrained representations. Following this paradigm, several studies adapt PEFT to video understanding by enhancing pretrained backbones with specialized temporal modules. For instance, AIM [51] introduces temporal attention modules into frozen CLIP encoders, whereas ST-Adapter [36] augments transformer blocks with lightweight spatio-temporal operators to facilitate video representation learning.

More recently, PEFT has also been explored in temporal action detection. AdaTAD [31] introduces lightweight adapters into a frozen VideoMAE backbone, enabling end-to-end optimization while preserving the efficiency advantages of parameter-efficient tuning. Despite their effectiveness, existing adapter-based methods predominantly perform spatial and temporal adaptation within a shared adaptation pathway. Such a design may be insufficient for micro-gesture recognition, where discriminative cues are characterized by subtle appearance variations and weak motion patterns. To better capture these fine-grained spatio-temporal characteristics, we propose a Spatial-Temporal Decoupled Adapter that explicitly separates spatial and temporal adaptation into dedicated branches, allowing each branch to focus on complementary aspects of micro-gesture representation learning.

2.3 Temporal Action Detection

Temporal Action Detection [26,31,52,53] aims to localize and classify action instances in untrimmed videos. Existing methods can be broadly categorized into three paradigms: two-stage approaches, single-stage methods, and end-to-end optimization. Early methods typically adopt a two-stage [26,32,54] paradigm, where temporal proposals are first generated and then classified using separate detection heads. While effective, such approaches suffer from information loss due to offline feature extraction and require multiple training stages.

Single-stage TAD methods have emerged to simplify the pipeline and improve efficiency. ActionFormer [52] introduces a multi-scale transformer architecture with local self-attention within each temporal scale and cross-scale feature fusion, enabling direct per-frame classification and boundary regression without explicit proposals. TriDet [42] enhances boundary precision via a trident-head design, while TemporalMaxer [44] demonstrates that simple max-pooling can replace complex temporal attention with competitive accuracy at lower computational cost. DyFADet [50] further leverages dynamic feature aggregation to adaptively fuse multi-scale temporal features based on content-dependent routing.

End-to-end approaches have also gained attention, particularly for scenarios with large pretrained backbones. PointTAD [43] adopts a sparse query-based paradigm with learnable keyframe points for multi-label temporal action detection using only RGB input. Building on this idea, AdaTAD [31] introduces lightweight temporal adapters into a frozen VideoMAE backbone, jointly optimizing the backbone and a multi-scale detection head in an end-to-end manner. This eliminates the conventional two-step pipeline of offline feature extraction followed by separate detection training, and substantially improves localization quality. AdaTAD++ [1] further explores spatial-temporal decoupling for scalable action detection, mainly through separated adapter training and high-resolution inference. Different from it, our method adopts a parallel dual-branch adapter within each backbone block, where spatial and temporal branches are jointly trained and their complementary adaptation features are fused for fine-grained micro-gesture recognition.

3 Methodology

3.1 Problem Definition

Micro-gesture online recognition can be formulated as follows: given a streaming video $\boldsymbol{V} \in \mathbb{R}^{T\times H\times W\times 3}$, where H and W denote the height and width of each frame and T is the number of frames processed so far, its micro-gesture annotations can be represented as $\Psi_g = \{\varphi_i = (t_s, t_e, c)\}_{i=1}^{N}$, where t_s, t_e, and c are the start time, end time, and category of micro-gesture instance φ_i, and N is the total number of ground-truth instances. The goal of online recognition is to predict a candidate proposal set $\Psi_p = \{\hat{\varphi}_i = (\hat{t}_s, \hat{t}_e, \hat{c}, s)\}_{i=1}^{M}$ to cover Ψ_g in a causal manner, where s denotes the confidence score.

3.2 Overall Architecture

We design a parameter-efficient adaptation framework that preserves the spatial representation capability of pretrained Vision Transformers while incorporating necessary temporal modeling through lightweight modules. Given an input video clip $\boldsymbol{X} \in \mathbb{R}^{T\times H\times W\times 3}$, a 3D patch embedding layer first converts it into a token sequence $\boldsymbol{Z} \in \mathbb{R}^{N\times C}$, where C is the embedding dimension and N is determined by the spatio-temporal resolution after patchification. This sequence is then fed into a pretrained backbone with positional encodings added to preserve spatio-temporal structure.

To enable task adaptation while retaining pretrained knowledge, we freeze all backbone parameters and insert Spatial-Temporal Decoupled Adapters after each transformer block. As illustrated in Fig. 1, these lightweight bottleneck structures independently model spatial appearance patterns and temporal dynamics, avoiding feature entanglement and improving discriminative ability. After feature extraction, the token sequence is passed through a projection layer while preserving its temporal structure. The projected features are subsequently processed by the detection head to jointly predict temporal boundaries and category labels for micro-gesture instances.

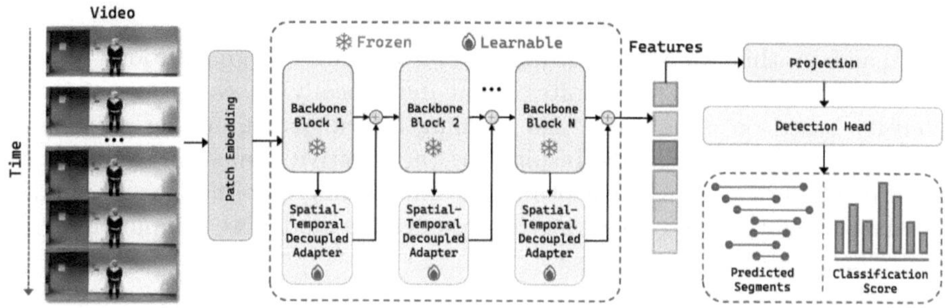

Fig. 1. Micro-gesture online recognition framework. The input video is patchified and processed by a frozen backbone. Trainable decoupled adapters separately model spatial and temporal information. Aggregated features are fed to the detection head for localization and classification.

3.3 Adaptive Soft Balanced Augmentation

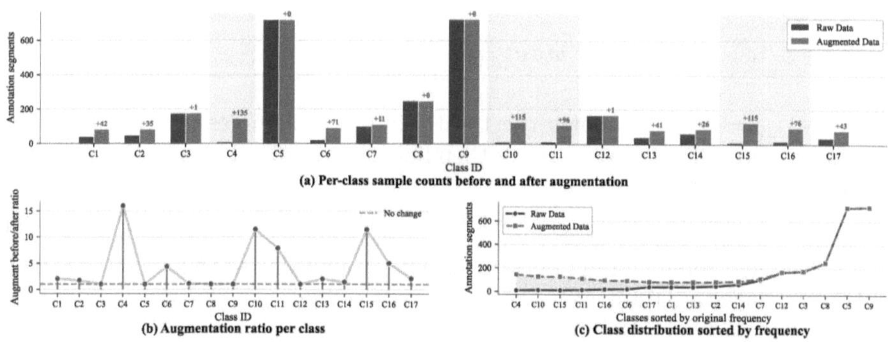

Fig. 2. Effect of ASBA on the class distribution of the SMG training set. (a) Raw vs augmented annotation segments. ASBA selectively boosts tail categories while keeping head classes nearly unchanged. (b) Class-wise augmentation ratios. Underrepresented classes receive stronger augmentation. (c) Sorted class frequency before and after augmentation. ASBA alleviates long-tailed imbalance while preserving the original data structure.

The SMG dataset exhibits a pronounced long-tailed distribution, where a few dominant categories contain significantly more annotation segments than the majority of tail classes. This causes models to be biased toward head classes and severely degrades recognition performance on rare micro-gesture categories. Conventional oversampling or fixed-threshold augmentation strategies fail to account for the varying degrees of underrepresentation across classes, often leading to overfitting on duplicated tail samples or insufficient augmentation for the most scarce categories.

To address this, we propose Adaptive Soft Balanced Augmentation (ASBA), a data-driven augmentation strategy that dynamically allocates augmentation intensity based on class rarity and learning difficulty, eliminating reliance on manually specified thresholds. Inspired by the effective number of samples proposed by Cui *et al.* [8], which captures the diminishing marginal benefit of additional data points through a geometric data overlap model, we compute an *effective sample count* for each class c:

$$E_c = \frac{1-\beta^{n_c}}{1-\beta}, \tag{1}$$

where n_c denotes the original sample count of class c, and β is a decay coefficient that controls the diminishing marginal contribution of additional samples.

Based on the effective sample count, we define a rarity weight:

$$w_r(c) = \frac{1/E_c}{\max_k(1/E_k)}, \tag{2}$$

this weight assigns larger values to underrepresented classes (Fig. 2).

To further account for class imbalance, we introduce a difficulty weight:

$$w_d(c) = \frac{N_{\text{total}} - n_c}{N_{\text{total}}}, \tag{3}$$

where $N_{\text{total}} = \sum_c n_c$ denotes the total number of samples across all classes. The rarity and difficulty weights are then linearly combined into a joint adaptive weight:

$$w_c = \lambda \cdot w_r(c) + (1-\lambda) \cdot w_d(c). \tag{4}$$

Rather than relying on a manually specified quantile as the balancing target, ASBA computes a data-driven *adaptive target ceiling*:

$$T^* = \frac{1}{|\mathcal{C}|} \sum_c n_c(1 + w_c). \tag{5}$$

For each tail class satisfying $n_c < T^*$, the augmented sample count is determined by:

$$\hat{n}_c = \left\lceil n_c + w_c \cdot \alpha \cdot \left(1 - \frac{n_c}{T^*}\right)(T^* - n_c)\right\rceil, \tag{6}$$

where α controls the overall augmentation intensity. The quadratic gap term ensures that the most underrepresented classes receive proportionally stronger augmentation, while head classes ($n_c \geq T^*$) remain unchanged.

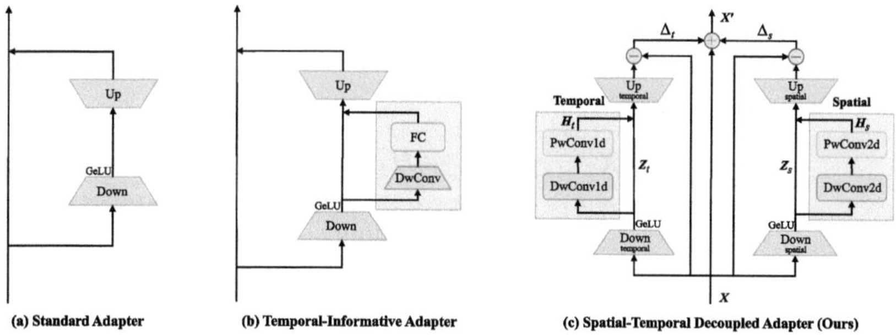

Fig. 3. Architectural comparison of adapter variants. (a) Standard Adapter: a bottleneck consisting of a down-projection, GeLU activation, and up-projection, with a residual connection. (b) Temporal-Informative Adapter: extends the standard design by inserting a depthwise convolution and a fully connected layer between the down- and up-projections. (c) Spatial-Temporal Decoupled Adapter (Ours): routes the input into parallel temporal and spatial branches, each performing depthwise and pointwise convolutions independently; the deltas from both branches are summed and added back to the input.

3.4 Spatial-Temporal Decoupled Adapter

Pretrained Vision Transformers encode rich spatial representations but lack temporal modeling capabilities. Existing adapter approaches often apply a single module that jointly processes spatial and temporal information, which may lead to feature entanglement. We argue that spatial appearance patterns and temporal dynamics should be modeled independently to avoid interference and enable more targeted adaptation.

As shown in (Fig. 3), Spatial-Temporal Decoupled Adapter consists of two parallel branches inserted after each Transformer block: a Temporal Adapter and a Spatial Adapter. Both branches follow a bottleneck design with down-projection, convolution, and up-projection.

Temporal Branch. Given input tokens $\boldsymbol{X} \in \mathbb{R}^{B_c \times N \times D}$, where B_c denotes the chunked batch size, N is the number of spatio-temporal tokens in each chunk, and D is the embedding dimension.

$$\boldsymbol{Z}_t = \mathrm{GELU}(\boldsymbol{X}\,\boldsymbol{W}^t_{down}), \tag{7}$$

where $\boldsymbol{W}^t_{down} \in \mathbb{R}^{D \times D'}$, $D' = \lfloor D \cdot r \rfloor$ with bottleneck ratio r. For temporal convolution, the hidden features are then reassembled into to $\boldsymbol{Z}_t \in \mathbb{R}^{b \cdot h \cdot w \times D' \times T}$, where b denotes the original batch size, T is the number of temporal positions, and h and w denote the spatial grid dimensions.

A depthwise 1D convolution is applied along the temporal axis for each spatial position:

$$\boldsymbol{H}_t = \mathrm{PWConv1D}(\mathrm{DWConv1D}(\boldsymbol{Z}_t)), \tag{8}$$

where DWConv1D denotes depthwise temporal convolution with kernel size k_t, and PWConv1D is a pointwise convolution for channel mixing. The output is reshaped back and projected up:

$$\boldsymbol{X}_t = \gamma_t \cdot (\boldsymbol{Z}_t + \boldsymbol{H}_t)\,\boldsymbol{W}^t_{up}, \tag{9}$$

where $\boldsymbol{W}^t_{up} \in \mathbb{R}^{D' \times D}$ is zero-initialized for training stability and γ_t is a learnable scale parameter.

Spatial Branch. The spatial branch shares the same bottleneck structure but applies depthwise 2D convolutions within each frame:

$$\boldsymbol{H}_s = \text{PWConv2D}(\text{DWConv2D}(\boldsymbol{Z}_s)), \tag{10}$$

where $\boldsymbol{Z}_s \in \mathbb{R}^{B_c \cdot t_1 \times D' \times h \times w}$ and $t_1 = N/(h \cdot w)$ denotes the number of temporal tokens per chunk. The spatial convolution captures local spatial patterns such as hand configurations and body part relationships:

$$\boldsymbol{X}_s = \gamma_s \cdot (\boldsymbol{Z}_s + \boldsymbol{H}_s)\,\boldsymbol{W}^s_{up}, \tag{11}$$

where $\boldsymbol{W}^s_{up} \in \mathbb{R}^{D' \times D}$ is the up-projection matrix of the spatial branch, also zero-initialized for training stability, and γ_s is a learnable scale parameter.

Parallel Combination. The two branches operate in parallel on the same input $\boldsymbol{X}$, and their outputs are combined additively:

$$\begin{aligned} \boldsymbol{X}' &= \boldsymbol{X} + (\boldsymbol{X}_t - \boldsymbol{X}) + (\boldsymbol{X}_s - \boldsymbol{X}) \\ &= \boldsymbol{X} + \Delta_t + \Delta_s, \end{aligned} \tag{12}$$

where $\boldsymbol{X}'$ is the adapted output that will be passed to the next Transformer block. The decoupled combination of the two branches ensures that each branch contributes an independent delta, preventing gradient interference between spatial and temporal learning signals.

4 Experiments

4.1 Experimental Setup

Datasets. The SMG dataset [6] contains 16 MG categories and one non-MG category, collected from 40 subjects. Following the official cross-subject protocol, 35 subjects are used for training and the remaining 5 for testing. Although both RGB and skeleton modalities are available, our method operates solely on RGB input.

Evaluation Metric. We perform experiments on the SMG dataset and adopt the F1 score and mAP as evaluation metrics. The formula of the F1 score is presented as follows:

$$\text{F1} = \frac{2 \cdot \text{Precision} \cdot \text{Recall}}{\text{Precision} + \text{Recall}}, \tag{13}$$

where Precision denotes the fraction of predicted micro-gestures that are correctly classified, and Recall denotes the fraction of ground-truth micro-gestures that are successfully detected. This metric captures both the temporal localization accuracy and the classification correctness in a single measure.

Implementation Details. We adopt VideoMAEv2-g [46] as our video backbone and enhance it with our Spatial-Temporal Decoupled Adapters for feature extraction. The adapter bottleneck ratio is set to 0.25, the temporal kernel size k_t is set to 3, and the spatial kernel size k_s is set to 3. ActionFormer head [52] is adopted as the detection head for temporal action localization. Videos are processed at the original frame rate of 28 fps with a sliding window of 16 frames and stride of 4 frames. All frames are resized to 160×160. We adopt the AdamW optimizer with a weight decay of 0.05. The learning rate is initialized to 1×10^{-4} and scheduled with cosine annealing over 400 epochs. The batch size is set to 4. For ASBA, we set $\beta = 0.999$, $\lambda = 0.7$, and $\alpha = 1.0$.

4.2 Experimental Results

Here, we evaluate our method on the SMG test set and compare with top-performing entries from both the current and previous challenge editions. As shown in Table 1, our method achieves an F1 score of 0.43808, ranking 1st in Track 2 of the 4th EI-MiGA-IJCAI Challenge, surpassing the second-place team by 2.926 percentage points. Compared with prior challenge winners, our approach substantially outperforms these methods, demonstrating substantial progress on this task.

4.3 Ablation Study

We conduct ablation studies to evaluate the contribution of each major design choice in our framework.

Scaling Backbone. To evaluate the impact of backbone scale, we compare VideoMAE-Small and VideoMAEv2-g in Table 2. Scaling up from the small to the giant variant raises the F1 score from 0.35614 to 0.41913, confirming that larger-scale pre-trained representations bring substantial benefits to micro-gesture recognition.

Table 1. The Results of Micro-gesture Online Recognition on the SMG test set. Data is provided by the Kaggle competition page[1].

Rank	Team	F1 Score
MiGA'26 1st	**XInsight Lab (Ours)**	**0.43808**
MiGA'26 2nd	AIM [27]	0.40882
MiGA'26 3rd	XD-L	0.35559
MiGA'25 1st	HFUT-VUT [29]	0.38027
MiGA'25 2nd	Chutian Meng [35]	0.31536
MiGA'24 1st	NPU-MUCIS [48]	0.27571
MiGA'24 2nd	HFUT-VUT [30]	0.14346
MiGA'23 1st	NPU-Stanford [14]	0.14810
MiGA'23 2nd	HFUT-VUT	0.04670

[1]The Kaggle competition page: https://www.kaggle.com/competitions/the-4th-ei-mi-ga-ijcai-challenge-track-2/leaderboard.

Component Contributions. To further understand the contribution of each component, we conduct an ablation study in Table 2. The default single-branch adapter in AdaTAD tends to entangle spatial and temporal features, limiting fine-grained discrimination. To mitigate the class imbalance issue, we apply data augmentation alone and obtain an F1 score improvement from 0.41913 to 0.42450, confirming its effectiveness in alleviating category imbalance. Furthermore, we integrate our proposed Spatial-Temporal Decoupled Adapter together with data augmentation. This combination increases the F1 score to 0.43808, indicating that decoupled spatial-temporal modeling provides a more effective adaptation mechanism for capturing subtle micro-gesture cues.

Augmentation Strategy. We evaluate the augmentation strategy on the SMG validation set in Table 3. Training without augmentation yields an average mAP of 25.10%, highlighting the severe impact of class imbalance. Fixed balancing targets based on mean or median statistics set a target sample size for each class as the mean or median of the class distribution and then augment minority classes up to that target. These targets offer some improvement, but their performance heavily depends on the choice of threshold. In contrast, ASBA achieves 29.05% average mAP, consistently outperforming both fixed-target baselines across all tIoU levels. This improvement can be attributed to ASBA's mechanism of adjusting augmentation intensity per class according to its effective sample count and learning difficulty, which avoids the over-augmentation or under-augmentation caused by a uniform threshold.

4.4 Error Analysis

In addition, we follow the standard practice in temporal action detection by adopting the diagnostic evaluation toolkit proposed by Alwassel *et al.* [2] to analyze model behavior from three perspectives. To better match the characteristics of the SMG dataset, we define Coverage groups as [XS, S, M, L, XL] with

Table 2. Ablation results of key modules on the SMG test set.

Data Augmentation	Spatial-Temporal Decoupled Adapter	Method	Backbone	F1 score
✗	✗	AdaTAD [31]	VideoMAE-S	0.35614
✗	✗	AdaTAD [31]	VideoMAE-B	0.38324
✗	✗	AdaTAD [31]	VideoMAEv2-g	0.41913
✓	✗	AdaTAD [31]	VideoMAEv2-g	0.42450
✓	✓	AdaTAD [31]	VideoMAEv2-g	**0.43808**

Table 3. Comparison of augmentation strategies under different tIoU thresholds on the SMG validation set.

Strategy	mAP									Avg
	@0.1	@0.2	@0.3	@0.4	@0.5	@0.6	@0.7	@0.8	@0.9	
None	36.16	35.95	35.13	33.48	31.33	25.61	19.61	7.22	1.43	25.10
Mean	38.62	38.30	37.64	34.34	32.52	29.29	24.94	15.27	1.42	28.04
Median	38.55	38.23	37.54	34.28	32.45	28.98	13.25	6.21	1.34	25.65
ASBA (Ours)	**39.04**	**38.74**	**38.28**	**35.75**	**33.60**	**30.65**	**25.55**	**17.72**	**2.08**	**29.05**

boundaries [0, 0.001, 0.0015, 0.0025, 0.0045, 1], Length groups as [XS, S, M, L, XL] with boundaries [0, 1, 1.5, 2.5, 4, INF], and Instance groups as [XS, S, M, L] with boundaries [-1, 50, 100, 200, INF].

False Negative Analysis. Figure 4 (a) and (b) compare the missed detection rates at tIoU $= 0.5$. Our method reduces false negatives primarily in medium-to-large coverage and longer-duration groups, indicating improved detection capability for temporally extended action instances. However, both methods still exhibit high miss rates on very short segments, suggesting that detecting brief micro-gestures remains challenging.

False Positive Analysis. Figure 4 (c) and (d) decompose false positive errors into five categories. Background error dominates the false positive composition in both methods and contributes the largest potential gain if removed. Compared to the baseline, our method slightly increases the true positive proportion in top-ranked predictions and reduces background errors, while localization and confusion errors remain comparable.

Sensitivity Analysis. Figure 4 (e) and (f) evaluate Average-mAP$_N$ across different characteristic groups. Our method improves the overall Average-mAP$_N$ from 26.4% to 32.7%, with particularly notable gains in large-coverage, longer-duration, and high-instance-count groups. The relative sensitivity analysis shows that our method achieves more balanced performance across different segment characteristics, with reduced sensitivity to coverage and length variations compared to the baseline.

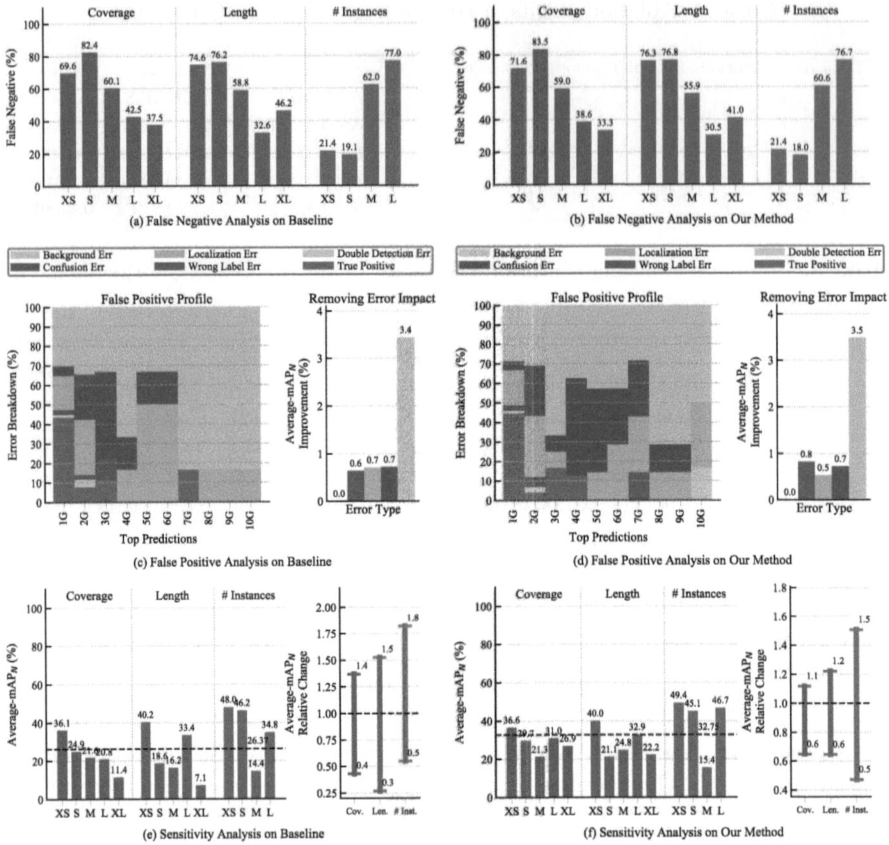

Fig. 4. Diagnostic evaluation of the baseline and our method on false negatives, false positives, and sensitivity analysis. The baseline adopts the AdaTAD model with VideoMAEv2-g as the backbone without data augmentation.

5 Conclusion

In this paper, we present our solution for the Micro-gesture Online Recognition track of the IJCAI 2026 MiGA Challenge. We propose a spatial-temporal-decoupled adapter together with an adaptive soft balanced augmentation strategy to model fine-grained spatial-temporal patterns and alleviate severe category imbalance. Specifically, the proposed adapter introduces parallel temporal and spatial branches into the frozen video backbone, capturing subtle motion dynamics and local appearance cues with few trainable parameters. The augmentation strategy dynamically adjusts augmentation intensity based on class rarity and sample distribution, improving tail-class representation while preserving head classes. Experimental results demonstrate that the proposed method achieves performance improvements by a large margin.

Although the proposed method achieves notable improvements, the accuracy of temporal localization and boundary detection remains limited. In future work, we plan to incorporate skeleton data, explore multi-modal fusion, and improve the detection head to further enhance temporal localization and recognition performance.

Acknowledgments. This work was supported by Anhui Provincial Natural Science Foundation (2408085J040), National Key R&D Program of China (2024YFB3311600), Natural Science Foundation of China (62272144, 72188101), the Major Project of Anhui Provincial Science and Technology Breakthrough Program (202423k09020001), and the New Cornerstone Science Foundation through the XPLORER PRIZE.

References

1. Agrawal, T., Ali, A., Dantcheva, A., Bremond, F.: Scaling action detection: AdaTAD++ with transformer-enhanced temporal-spatial adaptation. In: Proceedings of the IEEE/CVF International Conference on Computer Vision, pp. 12222–12231 (2025)
2. Alwassel, H., Caba Heilbron, F., Escorcia, V., Ghanem, B.: Diagnosing error in temporal action detectors. In: The European Conference on Computer Vision (2018)
3. Chen, G., et al.: Prototype learning for micro-gesture classification. arXiv preprint arXiv:2408.03097 (2024)
4. Chen, H., Liu, X., Li, X., Shi, H., Zhao, G.: Analyze spontaneous gestures for emotional stress state recognition: a micro-gesture dataset and analysis with deep learning. In: 2019 14th IEEE International Conference on Automatic Face & Gesture Recognition (FG 2019), pp. 1–8. IEEE (2019)
5. Chen, H., Schuller, B.W., Adeli, E., Zhao, G.: The 3rd challenge on human behavior analysis for emotion understanding (MiGA) 2025: from recognition to emotion understanding (2025)
6. Chen, H., Shi, H., Liu, X., Li, X., Zhao, G.: SMG: a micro-gesture dataset towards spontaneous body gestures for emotional stress state analysis. Int. J. Comput. Vision **131**(6), 1346–1366 (2023)
7. Chen, S., et al.: Adaptformer: adapting vision transformers for scalable visual recognition. In: Advances in Neural Information Processing Systems, vol. 35, pp. 16664–16678 (2022)
8. Cui, Y., Jia, M., Lin, T.Y., Song, Y., Belongie, S.: Class-balanced loss based on effective number of samples. In: Proceedings of the IEEE/CVF Conference on Computer Vision and Pattern Recognition, pp. 9268–9277 (2019)
9. Gao, R., Liu, X., Xing, B., Yu, Z., Schuller, B.W., Kälviäinen, H.: Identity-free artificial emotional intelligence via micro-gesture understanding. IEEE Trans. Affect. Comput. (2026)
10. Gu, J., et al.: Motion matters: motion-guided modulation network for skeleton-based micro-action recognition. In: Proceedings of the 33rd ACM International Conference on Multimedia, pp. 5461–5470 (2025)
11. Gu, J., Wang, F., Li, K., Wei, Y., Wu, Z., Guo, D.: MM-gesture: towards precise micro-gesture recognition through multimodal fusion. arXiv preprint arXiv:2507.08344 (2025)

12. Guo, D., Li, K., Hu, B., Zhang, Y., Wang, M.: Benchmarking micro-action recognition: dataset, methods, and applications. IEEE Trans. Circuits Syst. Video Technol. **34**(7), 6238–6252 (2024)
13. Guo, D., et al.: Mac 2024: micro-action analysis grand challenge. In: Proceedings of the 32nd ACM International Conference on Multimedia, pp. 11304–11305 (2024)
14. Guo, X., Peng, W., Huang, H., Xia, Z.: Micro-gesture online recognition with graph-convolution and multiscale transformers for long sequence. In: MiGA@ IJCAI (2023)
15. Hao, Y., et al.: Attention in attention: modeling context correlation for efficient video classification. IEEE Trans. Circuits Syst. Video Technol. **32**(10), 7120–7132 (2022)
16. Hao, Y., Zhang, H., Ngo, C.W., He, X.: Group contextualization for video recognition. In: Proceedings of the IEEE/CVF Conference on Computer Vision and Pattern Recognition, pp. 928–938 (2022)
17. Houlsby, N., et al.: Parameter-efficient transfer learning for NLP. In: International Conference on Machine Learning (2019)
18. Hu, E.J., et al.: Lora: low-rank adaptation of large language models. In: International Conference on Learning Representations (2022)
19. Jia, M., et al.: Visual prompt tuning. In: European Conference on Computer Vision (2022).
20. Li, K., Gu, J., Wang, F., Wu, Z., Fan, H., Guo, D.: Ma-bench: towards fine-grained micro-action understanding. arXiv preprint arXiv:2603.26586 (2026)
21. Li, K., et al.: Prototypical calibrating ambiguous samples for micro-action recognition. In: Proceedings of the AAAI Conference on Artificial Intelligence, vol. 39, pp. 4815–4823 (2025)
22. Li, K., Guo, D., Chen, G., Peng, X., Wang, M.: Joint skeletal and semantic embedding loss for micro-gesture classification. arXiv preprint arXiv:2307.10624 (2023)
23. Li, K., et al.: Mac 2025: the 2nd micro-action analysis grand challenge. In: Proceedings of the 33rd ACM International Conference on Multimedia, pp. 14216–14221 (2025)
24. Li, K., et al.: MMAD: multi-label micro-action detection in videos. In: Proceedings of the IEEE/CVF International Conference on Computer Vision, pp. 13225–13236 (2025)
25. Li, X.L., Liang, P.: Prefix-tuning: optimizing continuous prompts for generation. In: Proceedings of the 59th Annual Meeting of the Association for Computational Linguistics and the 11th International Joint Conference on Natural Language Processing (2021)
26. Lin, T., Liu, X., Li, X., Ding, E., Wen, S.: BMN: boundary-matching network for temporal action proposal generation. In: Proceedings of the IEEE/CVF International Conference on Computer Vision, pp. 3889–3898 (2019)
27. Liu, J., He, X., Liu, P., Shi, J., Zang, H., Hao, Y.: Motion reinforces appearance: RGB-skeleton gated residual fusion for micro-gesture online recognition. arXiv preprint arXiv:2606.11645 (2026)
28. Liu, P., et al.: A survey on fMRI-based brain decoding for reconstructing multimodal stimuli. arXiv preprint arXiv:2503.15978 (2025)
29. Liu, P., Li, K., Wang, F., Wei, Y., She, J., Guo, D.: Online micro-gesture recognition using data augmentation and spatial-temporal attention. arXiv preprint arXiv:2507.09512 (2025)
30. Liu, P., et al.: Micro-gesture online recognition using learnable query points. arXiv preprint arXiv:2407.04490 (2024)

31. Liu, S., Zhang, C.L., Zhao, C., Ghanem, B.: End-to-end temporal action detection with 1b parameters across 1000 frames. In: Proceedings of the IEEE/CVF Conference on Computer Vision and Pattern Recognition, pp. 18591–18601 (2024)
32. Liu, S., Zhao, X., Su, H., Hu, Z.: TSI: temporal scale invariant network for action proposal generation. In: Proceedings of the Asian Conference on Computer Vision (2020)
33. Liu, T., et al.: Self-supervised learning matters: a simple ensemble solution for micro-gesture recognition. arXiv preprint arXiv:2606.09261 (2026)
34. Liu, X., Shi, H., Chen, H., Yu, Z., Li, X., Zhao, G.: iMiGUE: an identity-free video dataset for micro-gesture understanding and emotion analysis. In: Proceedings of the IEEE/CVF Conference on Computer Vision and Pattern Recognition, pp. 10631–10642 (2021)
35. Meng, C., Ma, F., Zhang, C., Miao, J., Yang, Y., Zhuang, Y.: Online micro-gesture recognition in long videos via spatiotemporal feature encoding and query-based temporal detection. In: MiGA@ IJCAI (2025)
36. Pan, J., Lin, Z., Zhu, X., Shao, J., Li, H.: St-adapter: parameter-efficient image-to-video transfer learning. In: Advances in Neural Information Processing Systems, vol. 35, pp. 26462–26477 (2022)
37. Qian, W., et al.: Dual-path tokenlearner for remote photoplethysmography-based physiological measurement with facial videos. IEEE Trans. Comput. Soc. Syst. (2024)
38. Qian, W., Guo, D., Zhou, J., Zou, B., Yu, Z., Wang, M.: FreqPhys: repurposing implicit physiological frequency prior for robust remote photoplethysmography. arXiv preprint arXiv:2604.00534 (2026)
39. Qian, W., Li, K., Guo, D., Hu, B., Wang, M.: Cluster-Phys: facial clues clustering towards efficient remote physiological measurement. In: Proceedings of the 32nd ACM International Conference on Multimedia, pp. 330–339 (2024)
40. Qian, W., et al.: PhysDiff: physiology-based dynamicity disentangled diffusion model for remote physiological measurement. In: Proceedings of the AAAI Conference on Artificial Intelligence (AAAI) (2025)
41. Shang, T., Hao, Y., Pei, M., Li, K., Ben, H., Wang, S.: Cross-modal feature enhancement and contrastive alignment for micro-gesture recognition. In: Kittler, J., et al. (eds.) Chinese Conference on Pattern Recognition and Computer Vision (PRCV), pp. 203–217. Springer, Singapore (2025). https://doi.org/10.1007/978-981-95-5676-2_14
42. Shi, D., Zhong, Y., Cao, Q., Ma, L., Li, J., Tao, D.: TriDet: temporal action detection with relative boundary modeling. In: Proceedings of the IEEE/CVF Conference on Computer Vision and Pattern Recognition, pp. 18857–18866 (2023)
43. Tan, J., Zhao, X., Shi, X., Kang, B., Wang, L.: PointTAD: multi-label temporal action detection with learnable query points. In: Advances in Neural Information Processing Systems, vol. 35, pp. 15268–15280 (2022)
44. Tang, T.N., Kim, K., Sohn, K.: TemporalMaxer: maximize temporal context with only max pooling for temporal action localization. arXiv preprint arXiv:2303.09055 (2023)
45. Wang, C., Chen, H., Wei, H., Yang, Y., Chen, Y., Zhao, G.: iMigue-3k: a large-scale benchmark for micro-gesture analysis with self-supervised learning. arXiv preprint arXiv:2605.17179 (2026)
46. Wang, L., et al.: Videomae v2: scaling video masked autoencoders with dual masking. In: Proceedings of the IEEE/CVF Conference on Computer Vision and Pattern Recognition, pp. 14549–14560 (2023)

47. Wang, R., Li, K., Tong, A., Xu, J., Guo, D., Wang, M.: Gait emotion recognition via uncertainty-oriented class discriminative learning. IEEE Trans. Affect. Comput. 1–14 (2026)
48. Wang, Y., Kerui, L., Huang, H., Xia, Z.: Micro-gesture online recognition with dual-stream multi-scale transformer in long videos. In: MiGA@ IJCAI (2024)
49. Xia, Z., Huang, H., Chen, H., Feng, X., Zhao, G.: Hybrid-supervised hypergraph-enhanced transformer for micro-gesture based emotion recognition. IEEE Trans. Affect. Comput. (2025)
50. Yang, L., et al.: DyFADet: dynamic feature aggregation for temporal action detection. In: Leonardis, A., Ricci, E., Roth, S., Russakovsky, O., Sattler, T., Varol, G. (eds.) European conference on Computer Vision, pp. 305–322. Springer, Cham (2024). https://doi.org/10.1007/978-3-031-72952-2_18
51. Yang, T., Zhu, Y., Xie, Y., Zhang, A., Chen, C., Li, M.: Aim: adapting image models for efficient video action recognition. arXiv preprint arXiv:2302.03024 (2023)
52. Zhang, C.L., Wu, J., Li, Y.: ActionFormer: localizing moments of actions with transformers. In: Avidan, S., Brostow, G., Cissé, M., Farinella, G.M., Hassner, T. (eds.) European Conference on Computer Vision, pp. 492–510. Springer, Cham (2022). https://doi.org/10.1007/978-3-031-19772-
53. Zhao, C., Liu, S., Mangalam, K., Ghanem, B.: Re2tal: rewiring pretrained video backbones for reversible temporal action localization. In: Proceedings of the IEEE/CVF Conference on Computer Vision and Pattern Recognition, pp. 10637–10647 (2023)
54. Zhao, C., Thabet, A.K., Ghanem, B.: Video self-stitching graph network for temporal action localization. In: Proceedings of the IEEE/CVF International Conference on Computer Vision, pp. 13658–13667 (2021)

Motion Reinforces Appearance: RGB-Skeleton Gated Residual Fusion for Micro-gesture Online Recognition

Jialin Liu, Xinwen He, Pengyu Liu, Jiale Shi, Huaijuan Zang, and Yanbin Hao(✉)

School of Computer Science and Information Engineering, Hefei University of Technology (HFUT), Hefei, China
zanghj@hfut.edu.cn, haoyanbin@hotmail.com

Abstract. Micro-gesture analysis attracts increasing attention for inferring spontaneous emotion from subtle body movements. Micro-gesture online recognition, which localizes and classifies each gesture instance in untrimmed videos, is a core task in the 4th EI-MiGA-IJCAI Challenge. Compared with typical temporal action detection, MGR emphasizes the localization and classification of actions, requiring the model to output the start time, end time, and category of each micro-gesture. Moreover, since micro-gestures are highly spontaneous, relying solely on a single modality makes it difficult to capture the complete and accurate multi-modal cues. In this work, we propose **DyFADet+**, which extends DyFADet into a dual-stream RGB-skeleton framework. In our model, both modalities are projected into shared multi-scale temporal embeddings and fused through a gated residual module, which adaptively injects skeleton motion into the RGB representation rather than using naive concatenation. Finally these fused features are decoded by a Dynamic TAD head for online classification and boundary regression. On the SMG dataset, our method achieves an F1 score of 40.88, ranking 2nd in the Micro-gesture Online Recognition track.

Keywords: Micro-gesture online recognition · micro action · video understanding

1 Introduction

Body gestures are an important form of non-verbal communication and can reveal emotional states when facial expressions or speech are deliberately controlled. Subtle body movements are often harder to suppress and thus useful for inferring hidden affect [2,28]. Micro-gestures are brief, spontaneous movements, such as scratching the head or touching the nose, that arise under stress or discomfort, unlike intentional communicative gestures that accompany speech [5,13,14,28]. Understanding such cues supports applications in affective computing, human-computer interaction [32], and security-sensitive interview

H. Chen et al. (Eds.): IJCAI 2026, CCIS 3120, pp. 77–89, 2027.
https://doi.org/10.1007/978-981-92-4529-1_7

analysis [33], where the goal is to infer spontaneous or suppressed emotion from behavior rather than from explicit verbal reports [8,24].

Micro-gesture Online Recognition [4,31] requires methods to parse each untrimmed video into a set of gesture instances. For every detected segment, the method must predict both temporal boundaries and one of 17 fine-grained categories on the spontaneous Micro-Gesture dataset [5]. Unlike offline clip-level classification, MGR requires both temporal localization and fine-grained categorization within continuous, untrimmed videos. Most gesture and action recognition research still targets macro actions [7,18,42], and general temporal action detection methods are typically designed for longer, more salient motions with clearer boundaries [43,44]. Unlike macro actions, MGR faces unique challenges due to low-amplitude, weakly salient motions with fuzzy boundaries. Furthermore, micro-gestures are frequently obscured by irrelevant body movements [5,34]. While RGB streams retain rich appearance details, they often lack reliable motion cues for such fine-grained dynamics. Conversely, skeleton data provide distinct pose kinematics, but can be highly noisy due to the small displacements of micro-gestures [40]. Therefore, how to fully harness these two modalities to mitigate their respective weaknesses has become a critical challenge in MGR. Prior MGR methods explore dedicated skeleton-based graph convolution models [12], dual-stream Transformers [40], or query-based temporal designs [26]. Meanwhile, recent micro-gesture recognition methods investigate skeletal embeddings [19], prototype learning [3,17], and multimodal fusion [10]. However, how to fuse RGB and skeleton inside a unified temporal detector remains underexplored. In particular, naive early fusion or concatenation may amplify pose-estimation errors instead of exploiting cross-modal complementarity.

To address these issues, we extend DyFADet [43] into a dual-stream RGB-Skeleton architecture for MGR. Under this framework, RGB and skeleton inputs are first projected to a shared multi-scale temporal representation. In parallel, the RGB branch follows DyFADet's dynamic embedding projection, while the skeleton branch uses a convolutional Transformer projector with local temporal attention. At the fusion stage, a gated residual fusion module adaptively injects skeleton motion into the RGB features at each pyramid level. In particular, a lightweight adapter maps skeleton cues to the RGB channel space, and a sigmoid gate conditioned on both modalities decides where to trust kinematic information. Ultimately, the fused pyramid is decoded by the detection head for online classification and boundary regression. This design preserves discriminative appearance while selectively reinforcing motion-sensitive regions, effectively avoiding the instability of unconstrained concatenation.

The main contributions of this work are summarized as follows:

- We propose DyFADet+, a dual-stream RGB-Skeleton framework tailored for online micro-gesture recognition. By projecting both modalities into shared multi-scale temporal embeddings, our model effectively captures the short and subtle dynamics of micro-gestures.

- We design an innovative RGB-Skeleton Gated Residual Fusion module to leverage multi-modal cues. Instead of relying on naive concatenation, this module adaptively injects fine-grained skeleton motion into the RGB representation, providing precise features for the Dynamic TAD head to perform robust online classification and boundary regression.
- We validate our approach on the MiGA challenge benchmark[1], achieving F_1=40.88 and 2nd place on the leaderboard. Meanwhile, Extensive experiments on the SMG benchmark show that our method effectively enhances micro-gesture detection.

2 Related Work

2.1 Micro-gesture Datasets

Research on subtle body behaviour was driven by diverse benchmarks that differed in collection setting, annotation scales, and evaluation metrics [9,11,16,20]. For instance, iMiGUE [28] was a video dataset collected from many subjects and cultural backgrounds under realistic interview conditions. It provided dense frame-level micro-gesture annotations so that models had to infer suppressed or hidden affect from non-verbal cues rather than from subject identity or staged performance. SMG [5] focused on naturally occurring micro-gestures induced under psychological stress. It offered multimodal inputs including RGB and skeleton data, together with analyses linking gesture patterns to emotional stress. It defined 17 fine-grained categories and 40 samples. MA-52 [11] broadened the classification scheme to 52 whole-body micro-action categories with 7 body-part labels. These data were captured from face-to-face psychological interviews involving 205 participants and contained 22,422 trimmed clips for clip-level recognition with whole-body coverage. MMA-52 [20] further annotated untrimmed videos with start and end times and multiple co-occurring labels per interval, which formulated multi-label micro-action detection closer to temporal action detection than clip-level classification.

2.2 Temporal Action Detection

Micro-gesture online recognition coupled fine-grained classification with temporal localization in untrimmed videos [4,5]. Due to the subtle amplitude and brief duration of micro-gestures, their temporal boundaries remained inherently ambiguous. Furthermore, background body movements often co-occurred with target actions, which aligned the task closely with temporal action detection to predict precise intervals and category labels from continuous features. Existing methods handled these difficulties by capturing fine-grained structural and temporal context. Guo *et al.* [12] combined graph convolutions and multiscale

[1] The Kaggle competition page: https://www.kaggle.com/competitions/the-4th-ei-miga-ijcai-challenge-track-2/leaderboard.

Transformers to capture both joint structures and long-range dependencies. Wang *et al.* [40] proposed a dual-stream multiscale Transformer that processed RGB appearance and skeleton motion in parallel and fused multi-resolution temporal features for long interview videos. Liu *et al.* [26] introduced learnable query points that attend to informative temporal locations before classification and localisation, and further explored data augmentation with spatial-temporal attention for online recognition [25]. Yet, these methods emphasized specialized, modality-specific encoders rather than integrating strong pre-extracted RGB and skeleton descriptors inside a mature TAD decoder.

In parallel, general TAD methods that matured on benchmarks such as THUMOS-14 [15] and ActivityNet [1] became available in the OpenTAD toolbox [27]. Early milestones primarily relied on two-stage paradigms. For instance, BMN [22] generated complete candidate intervals by learning boundary-matching maps, while G-TAD [41] localized actions by matching sub-graphs constructed over video snippets. However, to bypass the complex and computationally expensive proposal generation process, the field subsequently shifted toward point-based architectures. Pioneering this shift, AFSD [21] exploited salient boundary cues for precise extent prediction without predefined reference windows. Building upon this, ActionFormer [44] significantly advanced the field by stacking transformer encoders with an FPN-style head to regress boundaries at multiple resolutions, and TriDet [36] further refined boundary quality via three-branch relative distance modeling. Despite these advances, managing complex action overlaps and high computational overhead remained challenging. To address these bottlenecks, PointTAD [38] adopted learnable query points for joint multi-label classification and localization. More recently, DyFADet [43] introduced dynamic feature aggregation across pyramid levels with a dynamic embedding projection and a DyHead decoder, successfully striking an optimal balance between detection performance and computational efficiency. Since direct concatenation may corrupt RGB features with noisy skeleton data, we avoid this naive method. Instead, we propose DyFADet+, which extends the original DyFADet into a dual-stream framework using a gated residual fusion module. This module adaptively injects skeleton motion cues into the RGB feature pyramid before the detection head. In this way, our approach keeps the RGB representation stable, allowing the network to automatically learn when and where the skeleton data is reliable.

2.3 Skeleton-Based Action Recognition

Skeleton sequences provide a clear way to represent joint movements without background interference. This makes them highly effective for micro-gesture analysis. Even when subtle hand or arm motions are obscured by cluttered backgrounds in RGB videos, they can still be accurately captured by the skeleton data. Early graph-based recognizers modeled the human skeleton as a spatio-temporal graph. Specifically, ST-GCN [42] applied graph convolutions with explicit spatial partitioning over joints and temporal neighborhoods. Subsequently, 2 s-AGCN [37] learned data-driven adjacency for joint and bone streams to capture action-specific structures. Furthermore, CTR-GCN [6] refined connectivity channel-wise so that different body parts attended to distinct relational patterns. Alternatively, PoseC3D [7] stacked joint heatmaps into 3D volumes processed by 3D CNNs. This approach combined convolutional locality with explicit temporal stacking, which achieved strong results on large-scale action benchmarks.

For micro-gesture online recognition, MiGA solutions typically embedded skeletons inside end-to-end temporal architectures, using either graph convolutions coupled with multiscale Transformers [12] or parallel RGB-skeleton Transformer encoders [40]. This ensured that joint structures and long-range temporal context were learned jointly with localization. Furthermore, data augmentation on pose sequences was shown to improve behavior analysis in multi-person settings [18]. However, such designs depended on pose estimates extracted from video and could be highly sensitive to estimation noise when movements were low-amplitude. In contrast, we treated the skeleton as a collaborative motion stream built from pretrained embeddings, projected them to the same multi-scale temporal pyramid as the RGB appearance, and integrated both through gated residual fusion before DyFADet decoding. This ensured that kinematic cues were selectively reinforced, rather than replaced, discriminative RGB features when pose information was reliable.

3 Method

3.1 Task Definition

We formulate micro-gesture online recognition as a temporal action detection problem on the SMG dataset [4,5]. Given an untrimmed video sequence $\mathcal{V}$, we represent it through a dual-stream feature representation to capture both appearance and kinematics. Specifically, the video is encoded into an RGB feature sequence $\mathbf{V}^r = \{\mathbf{v}_t^r\}_{t=1}^T \in \mathbb{R}^{T \times d_r}$ using pre-trained visual encoders (*e.g.*, VideoMAE [39]), alongside a temporally aligned skeleton feature sequence $\mathbf{V}^s = \{\mathbf{v}_t^s\}_{t=1}^T \in \mathbb{R}^{T \times d_s}$. The objective is to predict a set of action instances $\Psi = \{\psi_1, \psi_2, \ldots, \psi_N\}$, where N denotes the total number of predicted instances. Each instance $\psi_i = (t_i^s, t_i^e, c_i)$ is characterized by its start time t_i^s, end time t_i^e, and category label $c_i \in \mathcal{C}$, where $\mathcal{C}$ is the set of all predefined micro-gesture categories.

3.2 Pipeline

We build DyFADet+ on DyFADet [43], a dynamic feature aggregation detector for temporal action localization, and replace its single-stream projection with a two-stream RGB-Skeleton design. RGB appearance and skeleton motion are encoded into a shared multi-scale representation, fused with a gated residual module, normalized with the same lightweight multi-scale neck, and decoded by the dynamic detection head. The overall pipeline is illustrated in Fig. 1.

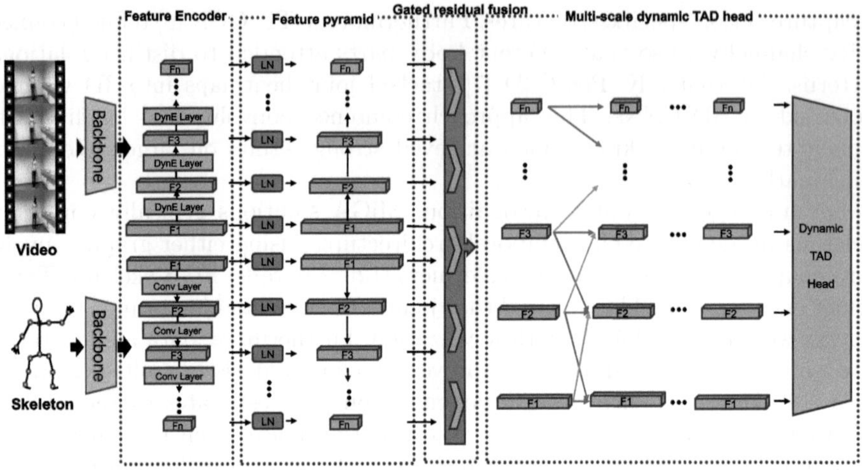

Fig. 1. Overview of the proposed DyFADet+ architecture for SMG micro-gesture detection.

3.3 RGB and Skeleton Feature

The RGB branch adopts the projection architecture of DyFADet [43], employing a shallow temporal convolutional embedder followed by a sequence of Dynamic Encoder (DynE) blocks. By leveraging dynamic feature aggregation and progressive temporal pooling, these blocks systematically downsample the sequence to construct a multi-scale feature pyramid $\{\mathbf{F}_r^{(\ell)}\}_{\ell=1}^{L}$. Parallel to the DynE blocks in the RGB branch, the skeleton branch employs a lightweight architecture driven by Conv blocks. We first embed the raw skeletal features into a higher-dimensional space using 1D convolutions. This is followed by initial Conv blocks that refine the representation at the original temporal resolution. A subsequent series of Conv blocks progressively downsamples the temporal dimension to construct a corresponding multi-scale pyramid $\{\mathbf{F}_s^{(\ell)}\}_{\ell=1}^{L}$ that is structurally aligned with the RGB branch. To guarantee temporal alignment prior to fusion, we apply nearest-neighbor interpolation to upsample the skeleton features whenever their temporal length is shorter than the corresponding RGB counterpart.

3.4 Gated Residual Fusion

Naive early fusion or concatenation can amplify noise when pose estimation is unstable. We instead inject skeleton cues as a gated residual on top of the RGB features at every pyramid level ℓ. A lightweight adapter $\psi_\ell(\cdot)$ maps the skeleton features to the RGB channel space, while a level-specific gate observes both

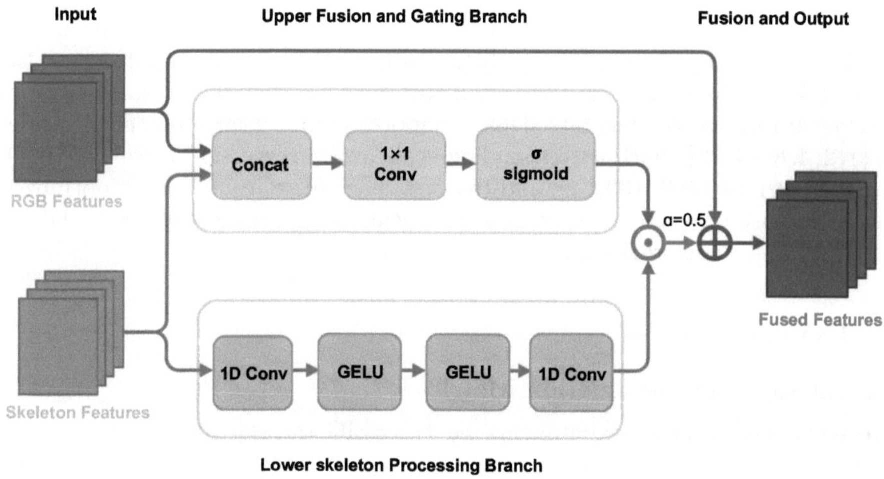

Fig. 2. Gated residual fusion at a single pyramid level: skeleton features pass through an adapter, a sigmoid gate (conditioned on RGB and skeleton) selects motion cues, and the result is added to RGB with scaling α.

modalities:

$$\mathbf{G}^{(\ell)} = \sigma\Big(\phi_\ell\Big(\Big[\mathbf{F}_r^{(\ell)}, \mathbf{F}_s^{(\ell)}\Big]\Big)\Big), \tag{1}$$

$$\mathbf{F}_f^{(\ell)} = \mathbf{F}_r^{(\ell)} + \alpha \cdot \mathbf{G}^{(\ell)} \odot \psi_\ell\Big(\mathbf{F}_s^{(\ell)}\Big), \tag{2}$$

where the bracketed term $[\mathbf{F}_r^{(\ell)}, \mathbf{F}_s^{(\ell)}]$ concatenates the RGB and skeleton features along the channel dimension (the per-time-step feature axis); ϕ_ℓ is implemented as a 1×1 temporal convolution, σ is the sigmoid function, $\odot$ denotes element-wise multiplication, and α=0.5 scales the residual branch. The adapter ψ_ℓ is a two-layer temporal convolutional stack with GELU activations. Each level ends with temporal layer normalization and dropout (p=0.05) on the skeleton residual. We initialize the bias of the gate convolution to -2 so that $\mathbf{G}^{(\ell)}$ starts near zero. This stabilizes early training. It forces the network to focus on the RGB path first, preventing noisy skeleton features from harming performance (Fig. 2).

3.5 Temporal Detection Head and Losses

Upon obtaining the fused multi-scale pyramid $\{\mathbf{F}_f^{(\ell)}\}_{\ell=1}^{L}$, the features are directly fed into the dynamic detection head [43]. Following the mechanism of DyFADet, the DyHead dynamically adjusts the shared parameters based on the input and selectively fuses cross-level features. The final predictions are yielded by two parallel branches across all levels, namely a classification module and a regression module. The classification module employs a 1D convolution followed by a Sigmoid function to predict the probabilities of the K=17 micro-gesture categories at each timestamp. Concurrently, the Reg module utilizes a 1D convolution with a ReLU activation to estimate the distances from the current timestamp to the start and end boundaries of the action instances. To handle micro-gestures of varying durations, we generate dense temporal points using predefined strides of $\{1, 2, 4, 8, 16, 32\}$, with corresponding scale-aware regression ranges of $(0, 4)$, $(4, 8)$, $(8, 16)$, $(16, 32)$, $(32, 64)$, and $(64, \infty)$. The entire framework is optimized end-to-end using the focal loss [23] for classification and the Distance-IoU (DIoU) loss [45] for segment regression.

4 Experiments

4.1 Dataset and Evaluation Metric

Dataset. We conduct experiments on the SMG dataset [5], which contains 3,692 annotated micro-gesture instances across 17 categories collected from 40 subjects. Crucially, the dataset provides RGB videos and skeleton sequences, making it an ideal benchmark for our proposed dual-stream architecture. The 40 subjects are divided into non-overlapping splits of 30 for training, 5 for validation, and 5 for testing.

Evaluation Metric. We jointly assess detection and classification using the F_1 score:

$$F_1 = \frac{2 \cdot \text{Precision} \cdot \text{Recall}}{\text{Precision} + \text{Recall}}. \tag{3}$$

For a long evaluation sequence, Precision is the fraction of correctly classified micro-gestures among all instances predicted by the model in that sequence, and Recall is the fraction of ground-truth micro-gestures that are both detected and correctly labeled. This metric comprehensively reflects the ability of a method to both detect and correctly classify micro-gestures.

4.2 Implementation Details

RGB features are extracted using VideoMAEv2-g [39], while the skeleton inputs utilize pre-computed kinematic representations. During training, we apply random temporal truncation. The feature stride is set to 4, and the sample stride to 1. The model is trained with AdamW [29] using an initial learning rate of 1e-4 and a weight decay of 0.025. We employ a cosine annealing learning rate schedule with a linear warm-up of 20 epochs, training for a total of 200 epochs. The training batch size is 2, and the validation and testing batch sizes are 1. Gradient clipping with a maximum norm of 1 and an exponential moving average (EMA) are applied.

Table 1. Leaderboard of the top submissions of the past four MiGA Track-2 editions (2023–2026) on the SMG online recognition track. Modalities: R=RGB, S=Skeleton.

Rank	Team	Core Methodology	Modality	F_1
MiGA'23 1st	NPU-Stanford [12]	Graph-Conv+Multiscale Transformer	S	14.81
MiGA'23 2nd	HFUT-VUT	ST-GCN	S	4.67
MiGA'24 1st	NPU-MUCIS [40]	PoseConv3D	R + S	27.57
MiGA'24 2nd	HFUT-VUT [26]	I3D+Mamba	R	14.35
MiGA'25 1st	HFUT-VUT [25]	VideoMAEv2-g + DyFADet	R	38.03
MiGA'25 2nd	Chutian Meng [30]	VideoMAE-g + DyFADet	R	31.54
MiGA'26 1st	XInsight Lab [35]	VideoMAEv2-g + AdaTAD	R	43.81
MiGA'26 3rd	XD-L	-	-	35.56
MiGA'26 2nd	**AIM (Ours)**	**VideoMAEv2-g + DyFADet**	**R + S**	**40.88**

4.3 Experimental Results

Table 1 summarizes the top-performing submissions across the past four editions of the MiGA challenge from 2023 to 2026 on the SMG online recognition track. The leaderboard highlights a clear evolutionary trajectory in micro-gesture online recognition. Early solutions primarily relied on skeleton-based architectures such as Graph-Conv and ST-GCN. In contrast, recent advancements are heavily driven by powerful foundational video models like VideoMAEv2-g. Notably, the state-of-the-art F_1 score has surged from 14.81 in 2023 to over 43.81 in 2026, reflecting the rapid progress in representation learning for fine-grained video understanding. In the 2026 edition, our AIM team secures the second place on the official leaderboard with an F_1 score of 40.88. By effectively integrating VideoMAEv2-g appearance features with structural skeleton cues through the DyFADet pipeline, our approach significantly outperforms the 2025 winning solution score of 38.03 by 2.85 points. The first-place submission by XInsight Lab [35] achieves 43.81 using an AdaTAD framework operating solely on RGB inputs. However, our dual-stream design combining RGB and skeleton demonstrates highly competitive performance. This result proves the continued viability and robustness of explicitly fusing joint kinematics with appearance representations for temporal micro-gesture localization.

4.4 Ablation Study

Table 2 shows the ablation results to evaluate different fusion methods. Using only RGB features achieves an F_1 score of 38.76, while using only skeleton features gets 22.31. This confirms that RGB appearance provides the main information for micro-gesture recognition. Directly concatenating the two modalities only improves the RGB baseline by 0.20, which means a simple merge cannot effectively use skeleton information. In contrast, our gated residual fusion achieves an F_1 score of 40.88. This is 2.12 higher than the RGB baseline and 1.92 higher than direct concatenation. These improvements prove that our learnable gate can successfully select useful skeleton cues and filter out noise, creating a more robust representation for detection.

Table 2. Ablation study of different modality integration strategies on the SMG test set.

ID	Variant	F_1
1	RGB only	38.76
2	Skeleton only	22.31
3	RGB + Skeleton	38.96
4	**Ours**	**40.88**

5 Conclusion

In this paper, we tackle the problem of micro-gesture online recognition, where naively fusing multi-modal cues often introduces disruptive noise. To address this, we propose DyFADet+, a novel dual-stream framework, as our solution for the MiGA challenge. Specifically, our method pairs RGB appearance features with skeleton motion descriptors. Rather than using conventional concatenation, DyFADet+ employs a gated residual module for multi-modal integration. This gating mechanism explicitly learns to selectively incorporate reliable skeleton kinematics while suppressing noisy pose estimates, yielding a highly robust multi-scale representation for temporal detection. Evaluated on the SMG benchmark, our method achieves an F_1 score of 40.88, securing 2nd place on the official leaderboard and outperforming last year's winning entry. Ablation studies further validate that our learnable cross-modal integration is superior to single-modality baselines and naive fusion. Future work will focus on stronger temporal modeling, long-tail learning, and effective multi-modal alignment to improve generalization on spontaneous micro-gestures.

Acknowledgments. This work was supported in part by the National Natural Science Foundation of China (No. 62472393), and by the Industry-University Cooperation Collaborative Education Project of the Ministry of Education (No. 250603873093026).

References

1. Caba Heilbron, F., Escorcia, V., Ghanem, B., Carlos Niebles, J.: ActivityNet: a large-scale video benchmark for human activity understanding. In: Proceedings of the IEEE/CVF Conference on Computer Vision and Pattern Recognition, pp. 961–970 (2015)
2. Calvo, R.A., D'Mello, S.: Affect detection: an interdisciplinary review of models, methods, and their applications. IEEE Trans. Affect. Comput. **1**(1), 18–37 (2010). https://doi.org/10.1109/T-AFFC.2010.1
3. Chen, G., et al.: Prototype learning for micro-gesture classification. arXiv preprint arXiv:2408.03097 (2024)

4. Chen, H., Schuller, B.W., Adeli, E., Zhao, G.: The 2nd challenge on micro-gesture analysis for hidden emotion understanding (MiGA) 2024: dataset and results. In: Proceedings of the IJCAI 2024 Workshop on Micro-gesture Analysis for Hidden Emotion Understanding (MiGA 2024). CEUR Workshop Proceedings, vol. 3848 (2024)
5. Chen, H., Shi, H., Liu, X., Li, X., Zhao, G.: SMG: a micro-gesture dataset towards spontaneous body gestures for emotional stress state analysis. Int. J. Comput. Vision **131**(5), 1346–1366 (2023). https://doi.org/10.1007/s11263-023-01761-6
6. Chen, Y., Zhang, Z., Yuan, C., Li, B., Deng, Y., Hu, W.: Channel-wise topology refinement graph convolution for skeleton-based action recognition. In: Proceedings of the IEEE/CVF International Conference on Computer Vision, pp. 13359–13368 (2021)
7. Duan, H., Zhao, Y., Chen, K., Lin, D., Dai, B.: Revisiting skeleton-based action recognition. In: Proceedings of the IEEE/CVF Conference on Computer Vision and Pattern Recognition, pp. 2969–2978 (2022)
8. Ekman, P., Friesen, W.V.: Nonverbal leakage and clues to deception. PsycEXTRA Dataset (1969). https://doi.org/10.1037/e525532009-012
9. Gu, J., et al.: Motion matters: motion-guided modulation network for skeleton-based micro-action recognition. In: Proceedings of the 33rd ACM International Conference on Multimedia, pp. 5461–5470 (2025)
10. Gu, J., Wang, F., Li, K., Wei, Y., Wu, Z., Guo, D.: MM-gesture: towards precise micro-gesture recognition through multimodal fusion. arXiv preprint arXiv:2507.08344 (2025)
11. Guo, D., Li, K., Hu, B., Zhang, Y., Wang, M.: Benchmarking micro-action recognition: Dataset, methods, and applications. In: IEEE Transactions on Circuits and Systems for Video Technology (2024). arXiv:2403.05234
12. Guo, X.P., Peng, W., Huang, H., Xia, Z.: Micro-gesture online recognition with graph-convolution and multiscale transformers for long sequence. In: Proceedings of the IJCAI 2023 Workshop on Micro-gesture Analysis for Hidden Emotion Understanding (MiGA 2023). CEUR Workshop Proceedings, vol. 3522 (2023)
13. Hao, Y., et al.: Attention in attention: modeling context correlation for efficient video classification. IEEE Trans. Circuits Syst. Video Technol. **32**(10), 7120–7132 (2022)
14. Hao, Y., Zhang, H., Ngo, C.W., He, X.: Group contextualization for video recognition. In: Proceedings of the IEEE/CVF Conference on Computer Vision and Pattern Recognition, pp. 928–938 (2022)
15. Idrees, H., et al.: THUMOS challenge: action recognition with a large number of classes. In: ECCV Workshop on Action Recognition with a Large Number of Classes (2014)
16. Li, K., Gu, J., Wang, F., Wu, Z., Fan, H., Guo, D.: MA-bench: towards fine-grained micro-action understanding. arXiv preprint arXiv:2603.26586 (2026)
17. Li, K., et al.: Prototypical calibrating ambiguous samples for micro-action recognition. In: Proceedings of the AAAI Conference on Artificial Intelligence, vol. 39, pp. 4815–4823 (2025)
18. Li, K., Guo, D., Chen, G., Liu, F., Wang, M.: Data augmentation for human behavior analysis in multi-person conversations. In: Proceedings of the 31st ACM International Conference on Multimedia, pp. 9516–9520 (2023)
19. Li, K., Guo, D., Chen, G., Peng, X., Wang, M.: Joint skeletal and semantic embedding loss for micro-gesture classification. arXiv preprint arXiv:2307.10624 (2023)
20. Li, K., et al.: MMAD: multi-label micro-action detection in videos. In: Proceedings of the IEEE/CVF International Conference on Computer Vision (2025)

21. Lin, C., et al.: Learning salient boundary feature for anchor-free temporal action localization. In: Proceedings of the IEEE/CVF Conference on Computer Vision and Pattern Recognition, pp. 3320–3329 (2021)
22. Lin, T., Liu, X., Li, X., Ding, E., Wen, S.: BMN: boundary-matching network for temporal action proposal generation. In: Proceedings of the IEEE/CVF International Conference on Computer Vision, pp. 3889–3898 (2019)
23. Lin, T.Y., Goyal, P., Girshick, R., He, K., Dollár, P.: Focal loss for dense object detection. In: Proceedings of the IEEE International Conference on Computer Vision, pp. 2980–2988 (2017)
24. Liu, P., et al.: A survey on FMRI-based brain decoding for reconstructing multimodal stimuli. arXiv preprint arXiv:2503.15978 (2025)
25. Liu, P., Li, K., Wang, F., Wei, Y., She, J., Guo, D.: Online micro-gesture recognition using data augmentation and spatial-temporal attention. arXiv preprint arXiv:2507.09512 (2025)
26. Liu, P., et al.: Micro-gesture online recognition using learnable query points. In: Proceedings of the IJCAI 2024 Workshop on Micro-gesture Analysis for Hidden Emotion Understanding (MiGA 2024). CEUR Workshop Proceedings, vol. 3848 (2024)
27. Liu, S., et al.: OpenTAD: a unified framework and comprehensive study of temporal action detection. In: Proceedings of the IEEE/CVF Conference on Computer Vision and Pattern Recognition Workshops, pp. 2650–2660 (2025)
28. Liu, X., Shi, H., Chen, H., Yu, Z., Li, X., Zhao, G.: iMiGUE: an identity-free video dataset for micro-gesture understanding and emotion analysis. In: Proceedings of the IEEE/CVF Conference on Computer Vision and Pattern Recognition, pp. 10631–10642 (2021)
29. Loshchilov, I., Hutter, F.: Decoupled weight decay regularization. In: International Conference on Learning Representations (2019)
30. Meng, C., Ma, F., Zhang, C., Miao, J., Yang, Y., Zhuang, Y.: Online micro-gesture recognition in long videos via spatiotemporal feature encoding and query-based temporal detection. In: Proceedings of the IJCAI 2025 Workshop on Micro-gesture Analysis for Hidden Emotion Understanding (MiGA 2025) (2025)
31. MiGA Organizers: MiGA: micro-gesture analysis for hidden emotion understanding challenge. https://cv-ac.github.io/MiGA2/ (2024). Accessed 10 May 2026
32. Noroozi, F., Corneanu, C.A., Kaminska, D., Sapinski, T., Escalera, S., Anbarjafari, G.: Survey on emotional body gesture recognition. IEEE Trans. Affect. Comput. **12**, 505–523 (2021). https://doi.org/10.1109/taffc.2018.2874986
33. Poppe, R., van der Zee, S., Taylor, P.J., Anderson, R.J., Veltkamp, R.C.: Mining bodily cues to deception. J. Nonverbal Behav. **48**, 137–159 (2024). https://doi.org/10.1007/s10919-023-00450-9
34. Shang, T., Hao, Y., Pei, M., Li, K., Ben, H., Wang, S.: Cross-modal feature enhancement and contrastive alignment for micro-gesture recognition. In: Kittler, J., et al. (eds.) Chinese Conference on Pattern Recognition and Computer Vision (PRCV), pp. 203–217. Springer, Singapore (2025). https://doi.org/10.1007/978-981-95-5676-2_14
35. Shen, X., Li, K., Wang, F., Qian, W., Jiang, J., Guo, D.: Spatial-temporal decoupled adapter for micro-gesture online recognition. arXiv preprint arXiv:2606.07355 (2026)
36. Shi, D., Zhong, Y., Cao, Q., Ma, L., Li, J., Tao, D.: Tridet: temporal action detection with relative boundary modeling. In: Proceedings of the IEEE/CVF Conference on Computer Vision and Pattern Recognition, pp. 18857–18866 (2023)

37. Shi, L., Zhang, Y., Cheng, J., Lu, H.: Two-stream adaptive graph convolutional networks for skeleton-based action recognition. In: Proceedings of the IEEE/CVF Conference on Computer Vision and Pattern Recognition, pp. 12026–12035 (2019)
38. Tan, J., Zhao, X., Shi, X., Kang, B., Wang, L.: PointTAD: multi-label temporal action detection with learnable query points. In: Advances in Neural Information Processing Systems, vol. 35 (2022)
39. Tong, Z., Song, Y., Wang, J., Wang, L.: VideoMAE: masked autoencoders are data-efficient learners for self-supervised video pre-training. In: Advances in Neural Information Processing Systems, vol. 35, pp. 10078–10093 (2022)
40. Wang, Y., Linghu, K., Huang, H., Xia, Z.: Micro-gesture online recognition with dual-stream multi-scale transformer in long videos. In: Proceedings of the IJCAI 2024 Workshop on Micro-gesture Analysis for Hidden Emotion Understanding (MiGA 2024). CEUR Workshop Proceedings, vol. 3848 (2024)
41. Xu, M., Zhao, C., Rojas, D.S., Thabet, A., Ghanem, B.: G-TAD: sub-graph localization for temporal action detection. In: Proceedings of the IEEE/CVF Conference on Computer Vision and Pattern Recognition, pp. 10156–10165 (2020)
42. Yan, S., Xiong, Y., Lin, D.: Spatial temporal graph convolutional networks for skeleton-based action recognition. In: Proceedings of the AAAI Conference on Artificial Intelligence, vol. 32 (2018)
43. Yang, L., et al.: DyFADet: dynamic feature aggregation for temporal action detection. In: Leonardis, A., Ricci, E., Roth, S., Russakovsky, O., Sattler, T., Varol, G. (eds.) Proceedings of the European Conference on Computer Vision. Springer, Cham (2024). https://doi.org/10.1007/978-3-031-72952-2_18
44. Zhang, C.L., Wu, J., Li, Y.: Actionformer: localizing moments of actions with transformers. In: Proceedings of the European Conference on Computer Vision, pp. 492–510. Springer, Cham (2022). https://doi.org/10.1007/978-3-031-19772-7_29
45. Zheng, Z., Wang, P., Liu, W., Li, J., Ye, R., Ren, D.: Distance-IoU loss: faster and better learning for bounding box regression. In: Proceedings of the AAAI Conference on Artificial Intelligence, vol. 34, pp. 12993–13000 (2020)

A Lightweight Skeleton Candidate Refinement Framework for Track 2 Online Micro-gesture Recognition

Zhiyong Liu, Jingwen Xu, and Yunan Li(✉)

School of Computer Science and Technology, Xidian University, Xi'an, China
24039100320@stu.xidian.edu.cn, yunanli@xidian.edu.cn

Abstract. Track 2 of the 4th EI-MiGA-IJCAI Challenge studies multimodality-based online micro-gesture recognition from long video sequences. The official task requires systems to detect and classify spontaneous micro-gesture events that appear sparsely and in arbitrary order, and it allows skeleton, RGB or multimodal inputs. This paper presents our skeleton-only submission: a lightweight two-stage candidate refinement framework. The first stage trains a skeleton temporal detector to generate reliable base predictions from frame-level skeleton features. To compensate for missed gestures, a low-threshold candidate pool is further generated. Instead of directly merging all low-threshold candidates, we introduce a candidate-level Add-on MLP refiner that decides whether an extra candidate should be added to the final prediction set based on its confidence, duration and relationship with base predictions. On our validation split, the proposed add-on refinement improves the event-level F1 score to 0.5491. The final Kaggle test submission achieved a public leaderboard score of 0.35302. The results indicate that candidate-level refinement is an effective and low-cost strategy for improving skeleton-based online micro-gesture recognition under limited training data.

Keywords: Continuous micro-gesture recognition · Skeleton-based temporal detection · Candidate refinement · Add-on MLP

1 Introduction

Micro-gestures are short, subtle and often unconscious body movements that may reveal hidden affective or cognitive states [3,17]. The 4th EI-MiGA-IJCAI Challenge extends the MiGA benchmark to three tracks and organizes Track 2 as multimodality-based online micro-gesture recognition from long video sequences [8,9]. In this track, micro-gesture samples are not pre-segmented or temporally aligned; instead, they occur spontaneously in arbitrary orders and may coexist with incomplete gestures, transitions and irrelevant background body movements. This setting is closer to real human communication, but it

Z. Liu and J. Xu—These authors are co-first authors of the article.

H. Chen et al. (Eds.): IJCAI 2026, CCIS 3120, pp. 90–103, 2027.
https://doi.org/10.1007/978-981-92-4529-1_8

also introduces significant difficulties in temporal localization and fine-grained category recognition.

Compared with conventional action detection, continuous micro-gesture recognition has three major challenges. First, micro-gestures usually have short durations and low motion amplitudes, so their temporal boundaries are difficult to localize. Second, background frames dominate long sequences, making the model prone to high false positives or missed weak gestures. Third, the training set is limited, which makes it difficult to train large-scale video models robustly. Although RGB data contain rich appearance information, lightweight skeleton features remain attractive because they compactly describe body motion while reducing irrelevant visual background.

A single skeleton detector often faces a precision–recall trade-off. High thresholds produce reliable predictions but miss subtle gestures, while low thresholds recover more potential gestures but introduce many false positives. Although recent online micro-gesture recognition methods have achieved promising results through multimodal modeling and advanced temporal architectures, most existing studies primarily focus on improving feature representations and detector designs. Comparatively little attention has been paid to recovering missed gesture instances from low-confidence predictions. This issue is particularly important for skeleton-only systems, where subtle gestures often receive weak confidence scores and are easily discarded by fixed decision thresholds. Therefore, we investigate a candidate refinement strategy that explicitly exploits low-threshold predictions to improve recall without substantially increasing false positives. To address this problem, we adopt a two-stage candidate refinement strategy. The first-stage skeleton detector produces a stable set of base predictions. A second low-threshold candidate pool provides potential missed segments. Finally, a candidate-level Add-on MLP refiner selectively adds useful extra candidates into the final prediction set.

The main contributions of this work are summarized as follows:

1. We propose a lightweight two-stage skeleton candidate refinement framework for continuous micro-gesture online recognition.
2. We introduce an Add-on MLP module that learns whether a low-threshold extra candidate can complement the base prediction set.
3. The method improves validation event-level F1 to 0.5491 and achieves a Kaggle public leaderboard score of 0.35302 in Track 2 of the 4th EI-MiGA-IJCAI Challenge.

2 Related Works

2.1 Micro-gesture and Micro-expression Analysis

Micro-gesture analysis is closely related to hidden emotion understanding and subtle nonverbal behavior recognition. Earlier micro-expression datasets, such as CASME II and SAMM, demonstrated the importance of spontaneous and fine-grained affective movements [7,29]. Different from facial micro-expression analysis, body micro-gesture analysis focuses on subtle body movements, which may be less affected by facial identity information. The iMiGUE dataset introduced identity-free video data for micro-gesture understanding and emotion analysis [17], while the SMG dataset further provided spontaneous body micro-gesture annotations and benchmark protocols for classification, online recognition and stress-state analysis [5,24]. The MiGA challenges have promoted this direction by organizing standardized tracks for micro-gesture classification, online recognition and, in 2026, behavior-based emotion understanding [3,9,21].

2.2 Online Micro-gesture Recognition

Online micro-gesture recognition differs from isolated classification because the model must localize and classify short events from long untrimmed sequences. Recent challenge solutions have investigated multi-scale and multimodal designs. The dual-stream multi-scale Transformer method combines RGB and skeleton information for temporal spotting and achieved the first ranking in the MiGA2024 online recognition track [27]. Learnable query point based schemes have also been explored for online micro-gesture spotting [16]. Although these methods demonstrate the value of richer temporal representations, they usually require more complex feature extraction or multimodal training. Moreover, most existing approaches mainly focus on improving temporal representations and detector architectures, while relatively little attention has been paid to exploiting low-confidence predictions during inference. As a result, weak micro-gesture instances are often missed when conservative thresholds are adopted. Different from these methods, our work focuses on a lightweight skeleton-only pipeline and introduces candidate-level refinement to recover useful low-threshold candidates while controlling false positives.

2.3 Skeleton-Based Action Modeling

Skeleton representations have been widely used for human action recognition because they encode body motion compactly. ST-GCN first introduced graph convolution for skeleton-based action recognition [28]. Later methods improved spatial-temporal modeling through adaptive graph convolution and shift graph convolution [6,23], as well as disentangled graph convolution and transformer-style architectures [18,22]. Inspired by these skeleton-based studies, our first-stage detector uses frame-level skeleton features to model long-term temporal context while keeping the pipeline lightweight.

2.4 Temporal Action Detection and Video Modeling

Temporal action detection and segmentation provide useful ideas for continuous micro-gesture spotting. Structured Segment Networks, BSN, BMN and ActionFormer localize actions by modeling proposals, boundaries or temporal contexts [13,14,30,31]. Temporal convolutional architectures such as TCN and MS-TCN have also been widely used for dense temporal action segmentation [1,11]. In RGB video understanding, I3D, R(2+1)D, SlowFast, TSM and

VideoMAE provide strong spatio-temporal representations [2,10,12,25,26]. Proposal generation and proposal refinement are widely studied in temporal action detection. Methods such as BSN and BMN generate large candidate pools and subsequently rank or refine proposals to improve localization quality. Inspired by these ideas, we view low-threshold micro-gesture predictions as candidate proposals and introduce a lightweight refinement module to determine whether they should be added to the final prediction set. However, although the official Track 2 permits skeleton, RGB or both modalities, our final 0.35302 submission uses only skeleton features. Therefore, instead of training a heavy RGB or multimodal video backbone, we adopt a lightweight skeleton detector and use a second-stage Add-on MLP to refine candidate-level predictions.

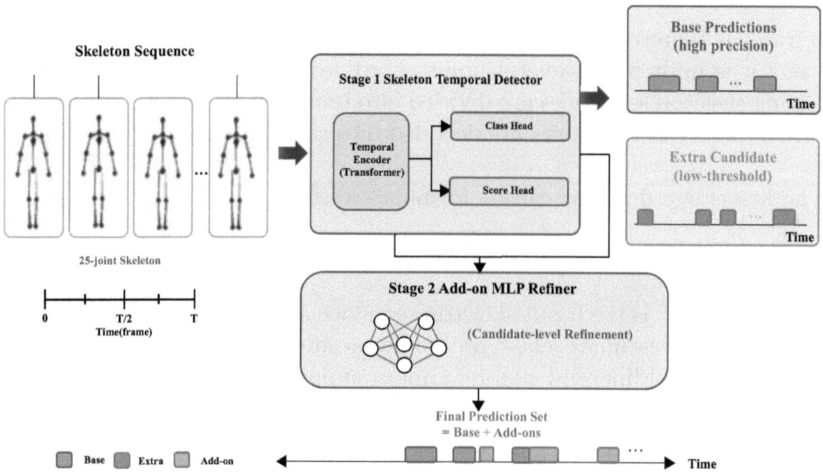

Fig. 1. Overview of the proposed two-stage skeleton candidate refinement framework for Track 2. A first-stage skeleton detector generates stable base predictions and a low-threshold candidate pool. The Add-on MLP refiner selectively adds useful extra candidates to obtain the final prediction set.

3 Proposed Method

3.1 Overall Framework

The overall framework is shown in Fig. 1. Given a long skeleton sequence from the Track 2 online recognition task, the first-stage temporal detector predicts candidate gesture segments. The detector output is decoded into two candidate sets: a high-confidence base prediction set and a low-threshold extra candidate pool. The base predictions provide stable results with relatively high precision, while the low-threshold pool is used to recover potential missed gestures.

Let the base prediction set be denoted as

$$\mathcal{B} = \{b_i = (c_i, s_i, e_i, p_i)\}_{i=1}^{N_b}, \tag{1}$$

where c_i, s_i, e_i and p_i denote the predicted class, start frame, end frame and confidence score. Similarly, the extra candidate set is

$$\mathcal{E} = \{e_j = (\hat{c}_j, \hat{s}_j, \hat{e}_j, \hat{p}_j)\}_{j=1}^{N_e}. \tag{2}$$

The goal of the second stage is to decide whether each extra candidate should be added to the final result set without introducing excessive false positives.

3.2 Skeleton Temporal Detector

Each frame is represented by a 525-dimensional skeleton feature vector. The feature vector is built from skeleton joint coordinates and motion-related descriptors. Long skeleton sequences are divided into temporal windows during training, and predictions from windows are decoded into candidate segments during inference.

The first-stage detector can be formulated as

$$\mathbf{H} = f_\theta(\mathbf{X}), \tag{3}$$

where $\mathbf{X} \in \mathbb{R}^{T\times 525}$ is the input skeleton sequence and f_θ is the temporal encoder. Prediction heads estimate class probabilities and temporal confidence scores. Class-wise thresholding and non-maximum suppression are used to obtain the base prediction set.

3.3 Low-Threshold Candidate Pool

The low-threshold candidate pool is generated by using a more permissive inference threshold. In our best setting, this pool corresponds to the internal score setting `s018`. It has higher recall potential but also contains many duplicate or low-quality segments. Directly merging all low-threshold candidates with the base predictions would increase false positives. Therefore, the pool is used only as a source of add-on candidates, and a candidate-level refiner is trained to select useful proposals.

3.4 Add-On MLP Refiner

The Add-on MLP is a lightweight candidate-level binary refiner. Unlike the first-stage detector, it does not process raw skeleton sequences directly. Instead, it operates on candidate metadata and overlap features. For an extra candidate e_j, we construct a feature vector

$$\mathbf{z}_j = \phi(e_j, \mathcal{B}), \tag{4}$$

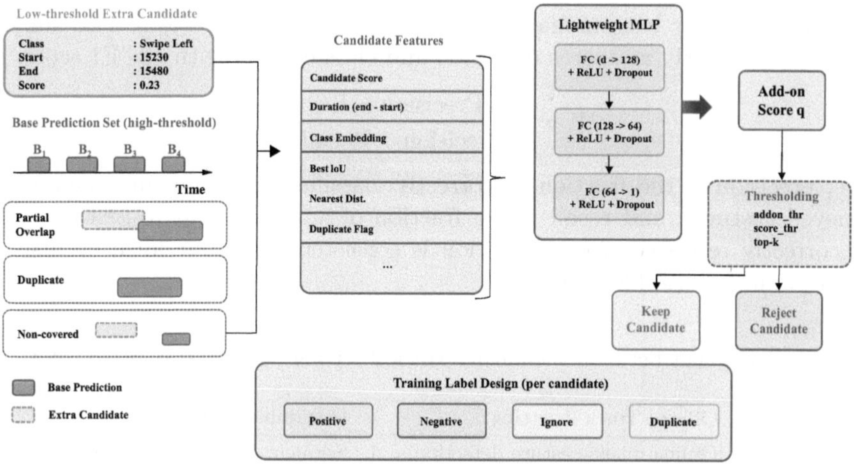

Fig. 2. Candidate-level Add-on MLP refiner. For each low-threshold candidate, the model uses its confidence, duration, class information and overlap relationship with base predictions to estimate whether the candidate should be added.

where $\phi(\cdot)$ includes the candidate confidence, temporal duration, predicted class and its relationship with nearby base predictions (Fig. 2).

The MLP outputs an add-on probability:

$$q_j = \sigma(g_\psi(\mathbf{z}_j)), \tag{5}$$

where g_ψ denotes the MLP and σ is the sigmoid function. The final decision is made by combining the add-on probability and the original candidate score. In our best validation setting, the selected parameters are `addon_thr` $= 0.50$, `extra_score_thr` $= 0.22$ and `topk` $= -1$.

During training, extra candidates are divided into four groups: positive add-on samples, negative samples, ignored ambiguous samples and base-covered duplicates. A positive add-on candidate is one that matches a ground-truth event but is not already covered by the base predictions. This design explicitly encourages the model to learn complementary candidates rather than simply re-detecting already covered events.

4 Experimental Setup

4.1 Challenge Task and Evaluation Protocol

Experiments are conducted on Track 2 of the 4th EI-MiGA-IJCAI Challenge [8, 9]. According to the official challenge description, Track 2 is multimodality-based online micro-gesture recognition from long video sequences. The benchmark is based on the SMG and iMiGUE datasets and allows participants to use skeleton data, RGB data or both modalities. The official evaluation uses a cross-subject protocol and jointly measures detection and classification with the F1 score:

$$F1 = \frac{2 \times \text{Precision} \times \text{Recall}}{\text{Precision} + \text{Recall}}. \tag{6}$$

Here, precision is the fraction of correctly classified micro-gestures among all retrieved gestures, and recall is the fraction of annotated micro-gestures that are correctly retrieved. This definition is consistent with our local event-level validation metric (Tables 1 and 2).

Table 1. Alignment between the official Track 2 setting and our submitted solution.

Item	Official Track 2 setting	Our implementation
Task	Online micro-gesture detection and classification in long sequences	Segment-level skeleton temporal detection with candidate refinement
Input modalities	Skeleton, RGB or multimodal input allowed	Skeleton-only submission
Evaluation	Detection and classification F1 under a cross-subject protocol	Event-level precision, recall and F1 for validation; Kaggle public score for test
Submission	Single CSV file on Kaggle	Final CSV generated by the Add-on MLP pipeline

4.2 Data and Local Validation Split

The official website states that the Track 2 training/validation portion contains long sequences from iMiGUE and SMG, while the test portion uses held-out iMiGUE long sequences. In this work, only skeleton data are used for the submitted 0.35302 solution. For local model selection, the labelled skeleton files available in our processed Phase1 package are split into 30 training samples and 5 validation samples. The training set contains 2452 annotated gesture segments, while the validation set contains 637 annotated segments. The processed Kaggle test subset used by our final script contains Sample0036–Sample0040 without public labels.

Table 2. Local split used for model development and ablation.

Split	Sample IDs	# Samples	# Segments
Train	Sample0001–Sample0030	30	2452
Validation	Sample0031–Sample0035	5	637
Kaggle test subset	Sample0036–Sample0040	5	–

4.3 Implementation Details

Our implementation is based on Python and PyTorch. The first-stage skeleton detector uses 525-dimensional frame-level skeleton features. The detector is trained on Phase1 training samples and validated on the held-out validation samples. The Add-on MLP is trained using base predictions and low-threshold `s018` candidates. We use event-level F1 on the validation set to select the final add-on threshold parameters.

The main files used in our best submission are summarized as follows:

- Base predictions: `outputs/skeleton_v1/*_class_thr_v1_raw.csv`;
- Low-threshold candidates: `outputs/stage2_refiner/*_s018_raw.csv`;
- Add-on refiner: `scripts/stage2_addon_refiner.py`;
- Kaggle output: `outputs/stage2_refiner/phase2_addon_mlp_v1_submission.csv`.

5 Results and Analysis

5.1 Validation Results

Table 3 shows the validation performance of different candidate refinement strategies. The first-stage skeleton baseline achieves an F1 score of 0.5151. A general candidate-level MLP refiner improves precision but reduces recall, resulting in a lower F1 score. In contrast, the proposed Add-on MLP improves recall while maintaining competitive precision, achieving the best validation F1 score of 0.5491.

Table 3. Validation event-level performance comparison.

Method	F1	P	R	TP	FP	FN	Rows
Skeleton v1 baseline	0.5151	0.6804	0.4144	264	124	373	388
Stage2 MLP refiner	0.5031	0.7159	0.3878	247	98	390	345
Proposed Add-on MLP	**0.5491**	0.6879	**0.4568**	**291**	132	**346**	423

5.2 Add-On Candidate Statistics

Table 4 reports the add-on training statistics. Most low-threshold candidates are already covered by the base predictions and are therefore treated as duplicates. Only a small number of candidates serve as true positive add-on samples. This observation explains why direct merging of low-threshold candidates is risky and why a candidate-level refiner is necessary (Fig. 3).

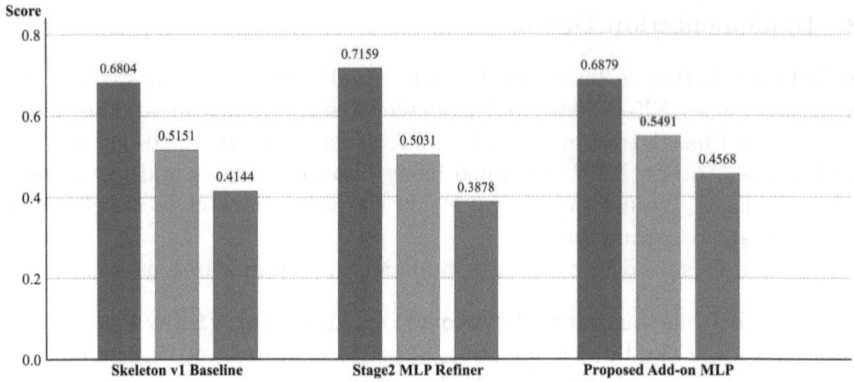

Fig. 3. Validation precision, recall and F1 comparison. The Add-on MLP improves the overall F1 mainly by increasing recall from the first-stage baseline.

Table 4. Add-on candidate label statistics.

Split	Positive	Negative	Ignore	Base-covered duplicate
Train	7	63	0	2017
Validation	24	68	10	343

5.3 Kaggle Test Submission

Using the selected validation parameters, the Add-on MLP is applied to the Kaggle test subset. The generated submission contains 432 predicted segments and achieves a public leaderboard score of 0.35302. The method is purely skeleton-based in this submission and does not use RGB video, optical flow, VideoMAE, SlowFast or other heavy video backbones (Table 5 and Fig. 4).

Table 5. Summary of the final submitted solution.

Item	Value
Modality	Skeleton only
First-stage detector	Skeleton temporal detector
Second-stage module	Add-on MLP candidate refiner
Validation F1	0.5491
Kaggle submission rows	432
Kaggle public score	**0.35302**

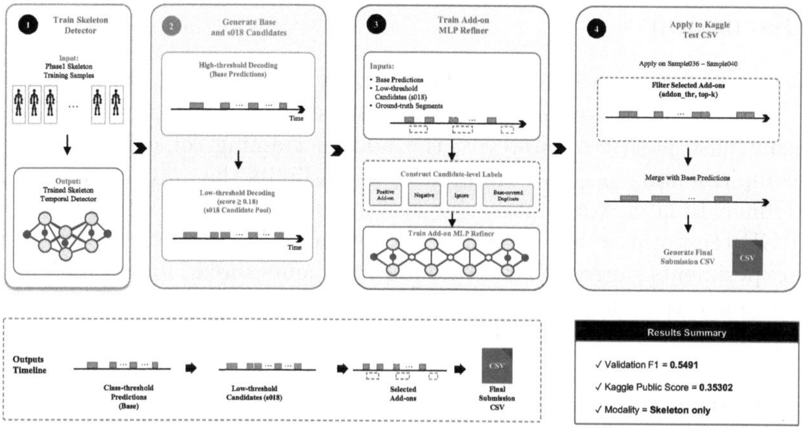

Fig. 4. Reproducible pipeline of the final Kaggle submission.

5.4 Comparison with Recent Challenge Results

To better position the proposed method, Table 6 compares our submitted solution with representative online micro-gesture recognition results reported in recent MiGA challenge editions [3,4,15,19,20]. The comparison is used for contextual positioning rather than as a strictly controlled benchmark, because the datasets, modalities, evaluation splits and challenge settings vary across years. Nevertheless, it provides a useful reference for understanding the competitiveness of the proposed lightweight skeleton-only submission.

Table 6. Contextual comparison with recent MiGA online micro-gesture recognition results.

Year	Representative method/team	Modality	Main strategy	F1 score
2023	HHuang	–	Online recognition leaderboard entry	0.1481
2024	NPU-MUCIS	RGB + skeleton	RGBPose-Conv3D + multiscale Transformer	0.2757
2025	HFUT-VUT	RGB	VideoMAEv2-g + DyFADet + STA	0.3803
2025	Chutian Meng et al.	RGB	VideoMAE-g + DyFADet	0.3153
2026	Ours	Skeleton only	Skeleton detector + Add-on MLP refinement	0.35302

As shown in Table 6, recent high-ranking systems usually rely on RGB video features, large pretrained video backbones or multimodal fusion. In contrast, our submission uses only skeleton features and a lightweight candidate-level refinement module. Although the scores are not directly comparable across different challenge years, the 2026 public leaderboard score of 0.35302 suggests that the proposed skeleton-only refinement strategy remains competitive among recent online micro-gesture recognition systems, especially considering its lower input and model complexity.

6 Discussion

The proposed approach is lightweight and easy to reproduce. Its improvement comes from exploiting complementary low-threshold candidates without heavily increasing false positives. However, the add-on training set contains very few truly complementary positive candidates, which limits the maximum gain of the MLP refiner. In later experiments, we also investigated RGB candidate verification, RGB temporal action detection and proposal-first skeleton segmentation. These experiments suggest that future improvements should focus on high-recall proposal generation and stronger segment-level classification, rather than simply lowering the detection threshold.

7 Conclusion

In this paper, we present a lightweight two-stage skeleton candidate refinement framework for continuous micro-gesture online recognition. The method first produces stable base predictions using a skeleton temporal detector and then selectively adds useful low-threshold candidates through an Add-on MLP refiner. Experiments show that the proposed strategy improves validation F1 to 0.5491 and achieves a Kaggle public leaderboard score of 0.35302. Future work will explore high-recall proposal generation, balanced hard-negative training and multimodal verification to further improve continuous micro-gesture detection.

Acknowledgments. We sincerely thank the organizers of the EI-MiGA-IJCAI Challenge for providing the benchmark platform, datasets, and evaluation infrastructure. We also thank the contributors of the SMG and iMiGUE datasets for advancing research in continuous micro-gesture understanding. In addition, we appreciate the valuable open-source resources and prior works from the research community that supported our implementation and experimentation. Finally, we thank our supervisors, teammates, and reviewers for their constructive suggestions and support throughout this work.

This work was jointly supported by the National Natural Science Foundation of China under Grant No. 62472342 and the Fundamental Research Funds for the Central Universities under Grant No. QTZX26117.

Disclosure of Interests. The authors have no competing interests to declare that are relevant to the content of this article.

Declaration on Generative AI. During the preparation of this work, the author(s) used AI-assisted writing tools for language polishing and structure organization. The author(s) reviewed and edited the content and take full responsibility for the publication's content.

References

1. Farha, A., Gall, Y., J.: MS-TCN: multi-stage temporal convolutional network for action segmentation. In: Proceedings of the IEEE/CVF Conference on Computer Vision and Pattern Recognition (CVPR), pp. 3575–3584 (2019)

2. Carreira, J., Zisserman, A.: Quo vadis, action recognition? A new model and the kinetics dataset. In: Proceedings of the IEEE Conference on Computer Vision and Pattern Recognition (CVPR), pp. 4724–4733 (2017). https://doi.org/10.1109/CVPR.2017.502
3. Chen, H., Schuller, B.W., Adeli, E., Zhao, G.: The 2nd challenge on micro-gesture analysis for hidden emotion understanding (MiGA) 2024: dataset and results. In: Proceedings of IJCAI 2024 Workshop & Challenge on Micro-gesture Analysis for Hidden Emotion Understanding (MiGA 2024). CEUR Workshop Proceedings, vol. 3848, pp. 1–9 (2024). https://ceur-ws.org/Vol-3848/paper_7.pdf
4. Chen, H., Schuller, B.W., Adeli, E., Zhao, G.: The 3rd challenge on human behavior analysis for emotion understanding (MiGA) 2025: from recognition to emotion understanding. In: Proceedings of IJCAI-2025 Workshop& Challenge on Human Behavior Analysis for Emotion Understanding (MiGA 2025). CEUR Workshop Proceedings, vol. 4168, pp. 1–7 (2025). https://ceur-ws.org/Vol-4168/paper_12.pdf
5. Chen, H., Shi, H., Liu, X., Li, X., Zhao, G.: SMG: a micro-gesture dataset towards spontaneous body gestures for emotional stress state analysis. Int. J. Comput. Vision **131**(6), 1346–1366 (2023). https://doi.org/10.1007/s11263-023-01761-6
6. Cheng, K., Zhang, Y., He, X., Chen, W., Cheng, J., Lu, H.: Skeleton-based action recognition with shift graph convolutional network. In: Proceedings of the IEEE/CVF Conference on Computer Vision and Pattern Recognition (CVPR), pp. 183–192 (2020)
7. Davison, A.K., Lansley, C., Costen, N., Tan, K., Yap, M.H.: SAMM: a spontaneous micro-facial movement dataset. IEEE Trans. Affect. Comput. **9**(1), 116–129 (2018). https://doi.org/10.1109/TAFFC.2016.2573832
8. EI-MiGA-IJCAI Challenge Organizers: The 4th EI-MiGA-IJCAI challenge track 2 (2026). https://www.kaggle.com/competitions/the-4th-ei-mi-ga-ijcai-challenge-track-2/overview. Accessed 12 May 2026
9. EI-MiGA Organizers: EI-MiGA-IJCAI2026: Human behavior analysis and interaction for emotional intelligence with the 4th MiGA challenge (2026). https://ei-miga.github.io/. Accessed 12 May 2026
10. Feichtenhofer, C., Fan, H., Malik, J., He, K.: Slowfast networks for video recognition. In: Proceedings of the IEEE/CVF International Conference on Computer Vision (ICCV), pp. 6202–6211 (2019). https://doi.org/10.1109/ICCV.2019.00630
11. Lea, C., Flynn, M.D., Vidal, R., Reiter, A., Hager, G.D.: Temporal convolutional networks for action segmentation and detection. In: Proceedings of the IEEE Conference on Computer Vision and Pattern Recognition (CVPR), pp. 1003–1012 (2017). https://doi.org/10.1109/CVPR.2017.113
12. Lin, J., Gan, C., Han, S.: TSM: temporal shift module for efficient video understanding. In: Proceedings of the IEEE/CVF International Conference on Computer Vision (ICCV), pp. 7083–7093 (2019)
13. Lin, T., Liu, X., Li, X., Ding, E., Wen, S.: BMN: boundary-matching network for temporal action proposal generation. In: Proceedings of the IEEE/CVF International Conference on Computer Vision (ICCV), pp. 3889–3898 (2019). https://doi.org/10.1109/ICCV.2019.00399
14. Lin, T., Zhao, X., Su, H., Wang, C., Yang, M.: BSN: boundary sensitive network for temporal action proposal generation. In: Ferrari, V., Hebert, M., Sminchisescu, C., Weiss, Y. (eds.) ECCV 2018. LNCS, vol. 11208, pp. 3–21. Springer, Cham (2018). https://doi.org/10.1007/978-3-030-01225-0_1

15. Liu, P., Li, K., Wang, F., Wei, Y., She, J., Guo, D.: Online micro-gesture recognition using data augmentation and spatial-temporal attention. In: Proceedings of IJCAI-2025 Workshop & Challenge on Human Behavior Analysis for Emotion Understanding (MiGA 2025). CEUR Workshop Proceedings, vol. 4168, pp. 1–11 (2025). https://ceur-ws.org/Vol-4168/paper_3.pdf
16. Liu, P., et al.: Micro-gesture online recognition using learnable query points. In: Proceedings of IJCAI 2024 Workshop & Challenge on Micro-gesture Analysis for Hidden Emotion Understanding (MiGA 2024). CEUR Workshop Proceedings, vol. 3848, pp. 1–10 (2024). https://ceur-ws.org/Vol-3848/paper_6.pdf
17. Liu, X., Shi, H., Chen, H., Yu, Z., Li, X., Zhao, G.: iMiGUE: an identity-free video dataset for micro-gesture understanding and emotion analysis. In: Proceedings of the IEEE/CVF Conference on Computer Vision and Pattern Recognition (CVPR), pp. 10626–10637 (2021). https://doi.org/10.1109/CVPR46437.2021.01049
18. Liu, Z., Zhang, H., Chen, Z., Wang, Z., Ouyang, W.: Disentangling and unifying graph convolutions for skeleton-based action recognition. In: Proceedings of the IEEE/CVF Conference on Computer Vision and Pattern Recognition (CVPR), pp. 143–152 (2020)
19. Meng, C., Ma, F., Zhang, C., Miao, J., Yang, Y., Zhuang, Y.: Online micro-gesture recognition in long videos via spatiotemporal feature encoding and query-based temporal detection. In: Proceedings of IJCAI-2025 Workshop & Challenge on Human Behavior Analysis for Emotion Understanding (MiGA 2025). CEUR Workshop Proceedings, vol. 4168, pp. 1–7 (2025). https://ceur-ws.org/Vol-4168/paper_4.pdf
20. MiGA-IJCAI Challenge Organizers: The 1st MiGA-IJCAI Challenge Track 2: Micro-gesture online recognition. CodaLab competition leaderboard (2023). https://codalab.lisn.upsaclay.fr/competitions/11782. Accessed 08 June 2026
21. MiGA Workshop Organizers: MiGA-IJCAI 2024 workshop and challenge: track 2 online micro-gesture recognition (2024). https://cv-ac.github.io/MiGA2/. Accessed 12 May 2026
22. Plizzari, C., Cannici, M., Matteucci, M.: Spatial temporal transformer network for skeleton-based action recognition. In: Del Bimbo, A., et al. (eds.) ICPR 2021. LNCS, vol. 12663, pp. 694–701. Springer, Cham (2021). https://doi.org/10.1007/978-3-030-68796-0_50
23. Shi, L., Zhang, Y., Cheng, J., Lu, H.: Two-stream adaptive graph convolutional networks for skeleton-based action recognition. In: Proceedings of the IEEE/CVF Conference on Computer Vision and Pattern Recognition (CVPR), pp. 12026–12035 (2019)
24. SMG Dataset Authors: SMG: spontaneous micro-gesture dataset and source code (2023). https://github.com/mikecheninoulu/SMG. Accessed 12 May 2026
25. Tong, Z., Song, Y., Wang, J., Wang, L.: VideoMAE: masked autoencoders are data-efficient learners for self-supervised video pre-training. In: Advances in Neural Information Processing Systems, vol. 35, pp. 10078–10093 (2022)
26. Tran, D., Wang, H., Torresani, L., Ray, J., LeCun, Y., Paluri, M.: A closer look at spatiotemporal convolutions for action recognition. In: Proceedings of the IEEE Conference on Computer Vision and Pattern Recognition (CVPR), pp. 6450–6459 (2018)
27. Wang, Y., Linghu, K., Huang, H., Xia, Z.: Micro-gesture online recognition with dual-stream multi-scale transformer in long videos. In: Proceedings of IJCAI 2024 Workshop & Challenge on Micro-Gesture Analysis for Hidden Emotion Understanding (MiGA 2024). CEUR Workshop Proceedings, vol. 3848, pp. 1–9 (2024). https://ceur-ws.org/Vol-3848/paper_2.pdf

28. Yan, S., Xiong, Y., Lin, D.: Spatial temporal graph convolutional networks for skeleton-based action recognition. In: Proceedings of the AAAI Conference on Artificial Intelligence, vol. 32, pp. 7444–7452 (2018)
29. Yan, W.J., et al.: CASME II: an improved spontaneous micro-expression database and the baseline evaluation. PLoS ONE **9**(1), e86041 (2014). https://doi.org/10.1371/journal.pone.0086041
30. Zhang, C., Wu, J., Li, Y.: Actionformer: Localizing moments of actions with transformers. In: Avidan, S., Brostow, G., Cissé, M., Farinella, G.M., Hassner, T. (eds.) Computer Vision – ECCV 2022, pp. 492–510. Springer, Cham (2022). https://doi.org/10.1007/978-3-031-19772-7_29
31. Zhao, Y., Xiong, Y., Wang, L., Wu, Z., Tang, X., Lin, D.: Temporal action detection with structured segment networks. In: Proceedings of the IEEE International Conference on Computer Vision (ICCV), pp. 2914–2923 (2017)

Rethinking the Role of Feature Engineering and Learning Strategies in Few-Shot Hidden Emotion Recognition

Xiaochuan Guo[1], Jihao Gu[2], Haixu Liu[3,4], Yuxin Liu[5], Qi Wang[6], Yufei Wang[7], Fei Wang[1,5](✉), Kun Li[8](✉), and Dan Guo[1,5]

[1] Hefei University of Technology, Hefei, China
jiafei127@gmail.com
[2] University College London, London, UK
[3] The University of Sydney, Sydney, Australia
[4] Beijing QBoson Quantum Technology Co., Ltd., Beijing, China
[5] Institute of Artificial Intelligence, Hefei Comprehensive National Science Center, Hefei, China
[6] Beihang University, Beijing, China
[7] The University of New South Wales, Sydney, Australia
[8] United Arab Emirates University, Abu Dhabi, United Arab Emirates
kunli.hfut@gmail.com

Abstract. In this paper, we present the solution developed by our team, XInsight Lab, which achieved first place in Track 3 of the 4th EI-MIGA-IJCAI Challenge with the test accuracy of 0.76923. To address the challenge of weak and sparse implicit emotion evidence in long videos, this paper extends the winning solution from the previous competition and proposes a compact multi-modal temporal modeling framework. The framework integrates and evaluates the effects of multi-source features, including 2D/3D skeletons, facial expression Blendshapes, DINOv2/v3 vision foundation models, X-CLIP video features, and Gemini semantic priors. Architecturally, we propose a cross-attention mechanism that utilizes static pose features (Base) as the Query and dynamic micro-motion differential features (Offset) as the Key/Value. By capturing local relative velocities, this mechanism eliminates static biases related to individual body shape and identity. Concurrently, an adaptive pooling method based on Multiple Instance Learning (MIL) is employed to extract instantaneous emotions while suppressing background noise in long sequences. Finally, the paper reveals the "representation collapse" phenomenon of general vision foundation models in micro-dynamic tasks, and analyzes the underlying mechanisms where networks fall into public-leaderboard-driven "pseudo-generalization" due to shortcut learning and rote memorization.

Keywords: Micro-gesture · Emotion Recognition · Multimodal Large Language Model

X. Guo, J. Gu, H. Liu, and Y. Liu—Equal contribution.

H. Chen et al. (Eds.): IJCAI 2026, CCIS 3120, pp. 104–121, 2027.
https://doi.org/10.1007/978-981-92-4529-1_9

1 Introduction

1.1 Background

Behavior-based hidden emotion understanding aims to infer latent affective states from subtle, spontaneous, and identity-insensitive nonverbal behaviors [5,11,33,42]. Different from conventional facial expression recognition or action recognition, the discriminative evidence is often weak, sparse, and distributed over long videos. In the iMiGUE setting [33], videos are muted post-match tennis interviews, and emotional polarity is inferred from the match outcome. This protocol reduces leakage from speech but makes the task strongly dependent on visual and behavioral cues. It also introduces three practical challenges: limited labeled data, class imbalance, and short-lived affective evidence that may only appear in a few frames.

Recent SOTA systems suggest that robust hidden emotion understanding requires complementary representations. Human-centric landmarks suppress background interference, self-supervised visual features encode appearance and context, and VLM-generated semantic descriptions provide high-level behavioral priors [9,37,39,43,44]. Prior challenge solutions further show the effectiveness of person cropping, whole-body pose features, temporal offsets, pseudo labels, and modality-wise training [13,50]. However, these methods leave an important modeling question underexplored: In the extreme setting of multimodal sparse features and few-shot supervision, what kind of feature engineering architecture and temporal learning strategy can genuinely drive the model to solidify robust representations with strong generalizability? Furthermore, within the weak-supervision iteration and voting ensemble paradigm relying on Vision-Language Models (VLMs), does the performance leap of downstream models stem from genuinely acquiring deep domain knowledge, or merely from memorizing noisy pseudo-labels, thereby achieving an empirical "pseudo-generalization" amid the prior randomness of the ensembled foundation models?

We address this question with a compact multimodal temporal framework. Our input contains body, face, and hand landmarks, facial expression descriptors, OpenPose whole-body keypoints, DINOv2/DINOv3 appearance features, X-CLIP video features, Gemini-based textual analyses encoded by BERT, optical-flow descriptors, and Depth Anything 3 (DA3) depth features [1,2,7,9,27,35–37,39,40]. For each temporal stream, we construct frame-wise offset features to capture small displacements that are often more discriminative than static coordinates or frame-level appearance alone, while optical flow and DA3 provide complementary motion and geometry cues.

Architecturally, to further investigate how base-offset features contribute to the model's overall performance, we explore two distinct fusion designs and a two-stage protocol. The first, *Independent*, regards original streams and offset streams as separate modalities and learns one temporal encoder per stream. The second, *Base-Offset Cross-Attention*, synchronously samples the base and offset sequences and uses base features as queries while using offsets as keys and values, following the Transformer attention formulation [41]. Each stream is encoded by

Transformer blocks with positional encoding, temporal consistency modeling, and attention pooling. The final multimodal representation is obtained by late concatenation and optimized with a lightweight MLP classifier. Meanwhile, we adopt a two-stage protocol: modality-wise pretraining followed by joint fine-tuning, augmented by iterative pseudo-labeling via voting ensembles for semi-supervised learning.

1.2 Contributions

Our contributions are threefold.

- **We propose** a novel perspective for hidden emotion recognition by reformulating it as offset-aware multimodal temporal modeling rather than generic video classification, which effectively guides the network to capture weak and sparse behavioral evidence.
- **We design** a rigorous controlled comparison between independent motion encoding and explicit base-offset cross-attention under identical feature and training protocols. Through this, we deeply investigate the underlying logic of performance gains and systematically expose critical pitfalls of "false improvements."
- **We demonstrate** the exceptional efficacy and superiority of our framework by achieving a leaderboard accuracy of 0.76923, ranking first in Track 3 of the 4th EI-MiGA-IJCAI Challenge.

2 Related Work

Data Preprocessing and Human-Centric Representations. Effective preprocessing is central to hidden emotion understanding because useful cues are often local, brief, and easily overwhelmed by background or camera variation. A common strategy is to convert raw videos into human-centric representations [22,32,38,45,48]. OpenPose [2] estimates whole-body keypoints with body, face, and hand landmarks using part affinity fields, while MediaPipe [35] provides efficient perception pipelines for body and hand tracking. Facial behavior toolkits such as OpenFace [1] further extract facial landmarks, head pose, gaze, and action-unit or expression descriptors. More recent whole-body pose estimators, including RTMPose, DWPose, ViTPose, and Sapiens, improve the speed, accuracy, or resolution of pose extraction through coordinate classification, two-stage distillation, plain vision Transformers, and large-scale human-centric pretraining [17,19,51,53]. These pose features reduce background noise but remain sensitive to occlusion, low-resolution hands, and head-pose changes.

Beyond coordinates, appearance and semantic features provide complementary evidence. Self-supervised image encoders such as DINOv2 and DINOv3 produce robust visual representations without task-specific fine-tuning [10,37,39,47], and video-language models such as X-CLIP extend image-language pretraining to temporal recognition [36]. VLMs can further generate structured behavioral descriptions that are encoded by language models such as BERT [7,9,12].

Another line of preprocessing reconstructs 3D facial or avatar geometry, e.g., FLAME-based faces and 3D Gaussian avatars, to obtain view-consistent deformation cues [18,26]. We adopt a lighter scalable alternative by extracting DA3-based monocular depth descriptors, which provide geometry-aware cues without explicit avatar reconstruction [27]. Optical flow is also used to encode short-term pixel-level motion, complementing landmark offsets that only describe motion on detected human keypoints [40]. In this work, we focus on a scalable 2D/appearance/motion/depth/semantic feature setting and explicitly augment temporal streams with frame-wise offsets when applicable.

Models for Micro-gesture and Hidden Emotion Understanding. Early video-based emotion or behavior recognition methods mainly rely on RGB backbones and generic temporal aggregation. I3D, TSM, and Video Swin Transformer model spatiotemporal appearance patterns from clips or frame sequences [3,28,34]. Skeleton-based methods such as ST-GCN and PoseConv3D replace pixels with structured keypoints and exploit spatiotemporal modeling over human joints [8,52]. While these architectures are effective for action recognition, hidden emotion understanding is harder because the relevant motion is subtle, non-periodic, and often not aligned with the anatomical adjacency used by graph models. Recent micro-gesture systems therefore move toward multi-stream modeling. RGB, skeleton, optical flow, depth, joint, limb, and textual streams have been combined through late fusion, cross-modal attention, prototype refinement, or modality-weighted ensembling [4,13,16,23,30,31,49]. Weak-to-Strong further shows that VLM-based pseudo labels, DINOv2 appearance features, and OpenPose offsets can improve multimodal hidden emotion recognition under limited supervision [50]. Regarding training strategies, existing systems often adopt temporal sampling, modality-wise pretraining, pseudo-label augmentation, focal or reweighted losses, and threshold or ratio calibration to achieve stable evaluation [6,29,50].

Benchmarks and Dataset Protocols. Several datasets have shaped the study of subtle behavioral cues. SMG [5] provides spontaneous micro-gesture videos for emotional stress state analysis. iMiGUE [33] focuses on identity-free micro-gesture understanding and emotion analysis from tennis interviews. More recently, iMiGUE-3K [42] scales micro-gesture analysis to thousands of long interview videos and supports foundation-model-style evaluation. Additionally, MA-52 [14], MMA-52 [25], and MA-Bench [20] belong to the broader field of subtle behavior analysis [11,21], their distinct target tasks make them complementary benchmarks for evaluating cross-task generalization. Their use in competitions [15,24] has also fostered broader community participation and helped drive progress in this field.

3 Methodology

3.1 Deep Feature Engineering and Representation Construction

To maximize the signal-to-noise ratio (SNR) of input video signals and explicitly expose extremely subtle motion cues within a high-dimensional feature space, this paper constructs a dimensionally rigorous and physically well-defined feature decoupling pipeline tailored for audiovisual signals in complex iMiGUE scenarios.

At the level of kinematic and dynamic feature extraction, we perform strict physical differentiation and parallel extraction of human limb and facial micro-movements across both the 2D pixel plane and 3D physical space. For 2D plane features, the OpenPose framework is utilized to densely extract absolute kinematic features based on the image coordinate system, which are then finely deconstructed into a 75-dimensional sequence of body landmarks, a 210-dimensional sequence of face landmarks, and a 126-dimensional sequence of hand landmarks. To capture 3D topologies with deep spatial structures, we concurrently employ the MediaPipe framework to extract 3D spatial coordinates containing depth priors, strictly reconstructing them into 198-dimensional 3D body landmarks, a high-fidelity 1434-dimensional dense 3D facial mesh vertex sequence, and 126-dimensional 3D hand landmarks. As shown in Fig. 1, we provide a visual example of 2D OpenPose keypoints and 3D MediaPipe-based spatial keypoint representations. Furthermore, to accurately and directly quantify localized facial muscle spasms and contractions, we independently derive 52-dimensional continuous facial expression blendshape parameters. To enable downstream networks to explicitly perceive the dynamic evolution of micro-expressions, for all the aforementioned 2D/3D skeleton and expression sequences, we retain the original parameters as "Base features" to represent static poses. Concurrently, we compute frame-by-frame numerical differentiation between adjacent valid frames along the temporal axis, explicitly constructing "Offset features" of identical dimensions. This operation essentially extracts the first-order derivative of human kinematics at the input level, forcing the model to shift its attention from absolute spatial poses—which suffer from static identity and body-shape biases—to localized relative micro-movement velocities. At the visual appearance and macro-contextual feature level, to counteract the catastrophic background noise in interview scenarios, we deploy the YOLOv11 object detection algorithm to perform fine, frame-by-frame cropping of the human subject. Subsequently, the purified human image sequences, stripped of background interference, are fed into DINOv2 to extract 1024-dimensional high-fidelity frame-level visual features. Concurrently, we employ an improved YOLOv26 combined with DINOv3 to extract a second set of fine-grained visual representations up to 4096 dimensions. This bounding-box-based spatial hard attention mechanism thoroughly prevents the network from overfitting to irrelevant backgrounds, while the dense receptive fields of the vision foundation models drastically enhance the network's generalization sensitivity toward localized subtle skin movement textures. Moreover, to compensate for the macro-scene contextual semantics lost during video cropping, we utilize a pre-trained X-CLIP model to extract 512-dimensional clip-level global semantic features from the uncropped original video, ultimately forming a complementary visual representation of "microscopic pure textures and macroscopic global context."

At the semantic prior level, purely data-driven deep vision networks often lack the psychological common sense required to infer human deceptive or masking

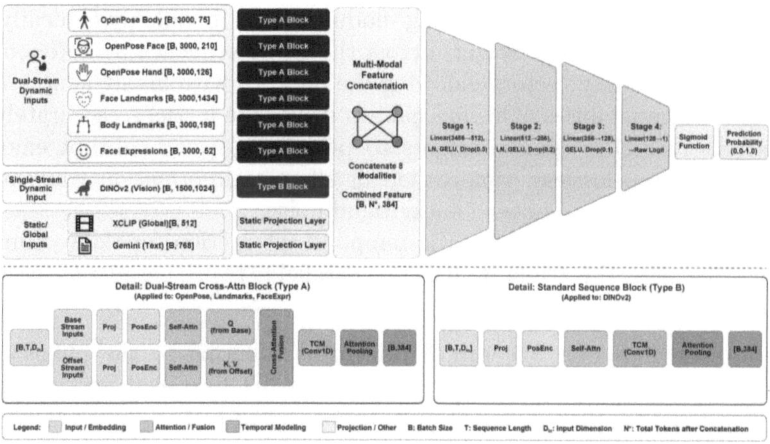

Fig. 1. Comparison of keypoint detection between MediaPipe and OpenPose.

behaviors. To address this, we introduce a Vision-Language Large Model (Gemini) to perform high-level semantic cognitive reasoning on video frame sequences. By combining advanced prompting strategies of "Chain-of-Thought and Reflection" (CoT+Reflection), we compel the large model to objectively extract Facial Action Coding System Action Units (FACS AUs), followed by logical deduction and self-reflection based on injected psychological prior rules. The generated structured analysis text is subsequently mapped into a 768-dimensional dense semantic feature tensor via the Qwen Embedding text encoder, thereby reducing the dimensionality of invaluable human cognitive logic into continuous variables that can be directly fused by downstream joint networks. Additionally, this paper explores the extraction of deformation offsets of dynamic 3D Gaussian point clouds in a canonical pose space based on the HRAvatar reconstruction algorithm as exploratory geometric topological features. However, extensive empirical validation reveals that certain samples in-the-wild datasets exhibit extreme rapid camera movements and severe physical occlusions, which easily trigger topological divergence in the underlying 3D reconstruction. To guarantee the absolute inferential robustness of the multimodal engineering system in unknown real-world environments, this exploratory geometric branch was proactively discarded in the final ensemble model.

3.2 Temporal Aggregation and Multimodal Network Architecture

Faced with the dozens of heterogeneous temporal feature sequences spanning thousands of frames extracted above, traditional global average pooling would mathematically flatten the abrupt emotional leakage signals occurring across just a few frames into the thousands of neutral background frames in terms of mathematical expectation. To circumvent this, we comprehensively adapt an adaptive attention pooling layer based on Multiple Instance Learning (MIL) at the end of all single-branch temporal encoding modules. This module dynamically computes frame-level attention weights across the entire sequence by autonomously learning a global query vector, endowing the network with highly adaptive temporal retrieval capabilities. This mechanism drives the model to accurately allocate extremely high response weights to peak frames with abrupt emotional changes, achieving lossless compression of effective information while aggressively suppressing long-sequence background noise.

Regarding the cross-modal spatio-temporal interaction topology of multidimensional features, this paper designs two parallel network paradigms for ensemble integration.

Independent Feature Stream Modeling: Adhering to the philosophy of extreme feature isolation, we treat all 2D and 3D base features, along with their corresponding offset differential features, as completely independent input sources that do not interfere with one another. They are fed separately into dedicated Transformer encoding networks and MIL pooling layers, and are only subjected to high-dimensional feature concatenation and joint semantic alignment at the fully connected layer at the deepest end of the network. This late fusion strategy maximizes the decoupling of heterogeneous features, effectively preventing highly sparse skeleton displacement coordinates and highly dense visual tensors from triggering catastrophic manifold interference in shallow networks.

Base-Offset Cross-Attention Modeling: As illustrated in Fig. 2, for modalities with strict kinematic derivation relationships (such as the 1434-dimensional 3D absolute facial coordinates and their corresponding 1434-dimensional temporal differences), the base features representing static underlying physical constraints serve as the Query vector, while the offset features representing dynamic minute changes serve as the Key and Value vectors $(Key, Value)$. The two components complete dense cross-stream interactions within a cross-attention module. This network architecture explicitly injects kinematic synergistic priors at the topological level, empowering the model with high-level reasoning capabilities to dynamically and conditionally evaluate the significance of localized minute deformations under specific underlying physical constraints of the torso and face.

3.3 Progressive Weakly-Supervised and Iterative Training Strategy

To address the extreme scarcity of effective strong annotations and the naturally occurring long-tailed distribution imbalance in the iMiGUE dataset, this study employs Focal Loss at the global joint classifier terminal to dynamically adjust the gradient penalty weights of samples. In the specific hyperparameter configuration, we set the initial adjustment factors to $\alpha = 0.35$ and $\gamma = 3.0$; as the semi-supervised learning-based label iterations progress, γ is gradually decreased, and α is moderately adjusted. Distinct from conventional algorithms that excessively inflate the γ value to force the model to obsess over hard examples, the core motivation behind our parameter setting is that, in-the-wild surveillance datasets of

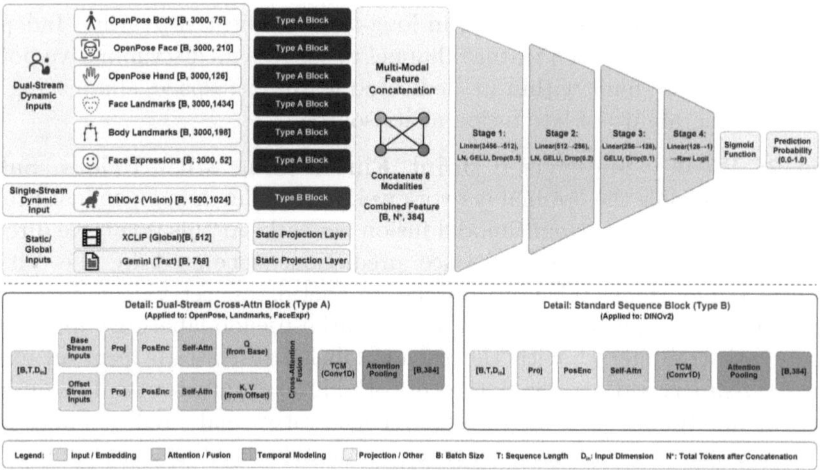

Fig. 2. Base-Offset Cross-Attention Model Structure.

concealed emotions, the "hard examples" in the model's eyes are often absolute noise and defective footage caused by severe camera defocusing, physical occlusions, or unavoidable subjective human annotation errors. This loss function parameter configuration explicitly prevents the model from over-focusing on invalid, extreme hard examples with exceptionally high losses. From a fundamental perspective, this substantially prevents the deep network from catastrophically memorizing and overfitting to anomalous noise under small-sample conditions, thereby substantially expanding the true generalization boundaries of the classification hyperplane on unseen test domains. To further optimize the model's predictive performance, we introduced a post-processing strategy based on a positive sample ratio threshold. By incorporating dynamic feedback from the leaderboard, parameter tuning was conducted within the range of 0.56–0.66 to determine the optimal threshold. The complete reproduction of the entire training pipeline relies on a single NVIDIA RTX Pro 6000 GPU, with a total computation time of approximately 3 days.

Based on the above settings, this paper executes a progressive, three-stage learning pipeline from shallow to deep:

Stage 1: Modality-Specific Independent Pre-training. At the beginning of model initialization, the system temporarily disconnects all high-level multimodal fusion channels. It relies solely on the limited training set samples with official ground-truth labels to conduct completely independent, purely supervised classification pre-training for each of the aforementioned individual 2D/3D skeleton channels and high-dimensional visual feature extraction branches. This strategy fundamentally circumvents the "modality inertia and competitive dominance" pitfalls during the joint initialization of complex multimodal networks, preventing the network from over-relying on the fast-converging dominant visual branches to rapidly minimize the overall loss, which would leave weak feature streams, such as motion differences, in long-term gradient starvation. Independent pre-training forces each feature channel to squeeze its discriminative potential to the absolute limit within its dedicated feature subspace, eliminating the weak-link vulnerability of late fusion at its source.

Stage 2: Weakly-Supervised Joint Fine-Tuning with Priors. Subsequently, we assemble the optimal network weights from each independent branch in Stage 1 to establish the multimodal fusion network. At this point, we directly extract the zero-human-cost inference predictions generated by the Vision-Language Large Model (Gemini) on the unlabeled test set videos during the feature engineering phase, treating them as initial pseudo-labels. By mixing the ground-truth training set with the pseudo-labeled test set to form an extensive weakly-supervised data pool, the network initiates global cross-domain joint alignment fine-tuning with an extremely conservative and small learning rate. This process not only expands the scale of available data but also compels the high-dimensional classification plane of the joint network to pre-fit and adapt to the vast and unknown underlying data distribution of the test set.

Stage 3: Iterative Self-training via Multi-model Voting Ensemble. As the multimodal joint network progressively converges on specific domain scenarios, its high-level discriminative capability in vertical emotion anti-spoofing tasks is further enhanced. To reduce noise in the initial pseudo-labels and elevate label quality, an iterative self-training mechanism is introduced in this stage. Rather than discarding the initial predictions generated by Gemini, we treat them as prior baselines, combining them with the topologies of multiple currently converged fine-tuned models to form an ensemble voting system. Subsequently, this batch of high-quality pseudo-labels is utilized to re-drive the independent modality training and joint fine-tuning of the previous two stages. The aforementioned process executes iteratively for several rounds, ultimately steering the decision boundary of the joint network toward stability and optimality.

4 Experiment

4.1 Ablation Study

This work extends last year's winning solution by introducing key enhancements. Through systematic ablation experiments, we identified effective training strategies and feature engineering methods tailored for foundation models. As shown in Table 1, each proposed component consistently improves performance over the baseline, with Multiple Instance Learning (MIL) contributing the most significant gain. Crucially, our results confirm that integrating Multi-Instance Learning (MIL) yields significant performance gains.

4.2 Contrast Experiment

This paper evaluates the performance of independent modality training versus the joint training of dual-architecture models under both supervised and semi-

Table 1. Ablation study of general enhancement strategies based on base-model

Method	ACC (%)	Δ (%)
MLP	63.46	–
+ Weak supervision	66.35	+2.89
+ Offset features	68.27	+1.92
+ Pre-training	69.23	+0.96
+ Multiple Instance Learning	70.19	+0.96

supervised learning frameworks. The experimental results align with expectations. First, the 3D-based MediaPipe demonstrates superior keypoint representation capabilities compared to the 2D OpenPose. As shown in Table 2, cross-modal joint fine-tuning benefits significantly from weak supervision and iterative pseudo-label refinement, achieving consistent improvements over independent training across all feature modalities. This advantage stems from the introduction of depth information via 3D coordinates, which enables a more accurate capture of spatial geometric structures and effectively mitigates viewpoint occlusion issues. Second, while the performance of the Base-Offset Cross-Attention mechanism is inferior to that of independent offset features during the independent pre-training phase, it reverses this trend and outperforms them during the joint fine-tuning phase. This is primarily because the cross-attention mechanism is highly susceptible to overfitting during independent pre-training; conversely, during joint fine-tuning, the multi-modal collaborative optimization and joint constraint mechanisms impose stronger generalization constraints on the model, thereby effectively mitigating this issue.

However, the experiments also expose several counter-intuitive anomalies, which will be analyzed in depth in the subsequent section.

5 Discussion

In this section, we move beyond specific quantitative metrics to conduct a deep qualitative analysis of several counter-intuitive phenomena observed during our experiments.

5.1 Macro-context and Semantic Priors Validity vs. Representation Collapse of Vision Foundation Models in Micro-dynamics

When exploring the effectiveness of features from different modalities, we observed an exceptionally pronounced divergence in efficacy between macro-features and micro-dense features. Our experiments demonstrate that X-CLIP video features and text features provided by the Vision-Language Model (Gemini) play a dominant, positive discriminative role within the joint network. Conversely, the dense feature pipeline combining YOLO (with fine target subject

Table 2. Performance of independent modality training and joint fine-tuning under multi-round weakly supervised training

Stage	Fusion	Feature	Supervised	Geminipseudo(acc 0.64)	Weaklypseudo(acc 0.69)	Weaklypseudo(acc 0.74)
Stage 1	Crossattention	Face expression	0.65	0.63	0.63	0.68
		Face landmark	0.64	0.62	0.61	0.63
		Body landmark	0.61	0.58	0.61	0.67
		Hand landmark	0.60	0.63	0.63	0.64
		OpenPose	0.60	0.62	0.63	0.62
	Independent	Face expression	0.64	0.63	0.63	0.63
		Face landmark	0.62	0.59	0.62	0.61
		Body landmark	0.66	0.58	0.60	0.65
		Hand landmark	0.64	0.63	0.63	0.68
		OpenPose	0.60	0.61	0.63	0.63
		Face expression offset	0.67	0.69	0.69	0.68
		Face landmark offset	0.63	0.65	0.64	0.63
		Body landmark offset	0.64	0.67	0.64	0.68
		Hand landmark offset	0.63	0.63	0.63	0.64
		OpenPose offset	0.63	0.59	0.63	0.64
		DINOv2	0.53	0.53	0.53	0.53
		DINOv3	0.53	0.53	0.53	0.53
		Gemini 3	0.63	0.66	0.70	0.72
		X-CLIP	0.63	0.65	0.71	0.74
Stage 2	Crossattention	—	0.63	0.70	0.73	0.77
	Independent	—	0.66	0.69	0.72	0.75

cropping) and DINO (v2/v3) yields virtually no substantial assistance to the classification boundary.

Delving into its physical and semantic essence, although X-CLIP sparsely samples only 8 frames out of a complete video sequence spanning thousands of frames—objectively losing millisecond-level micro-motion temporal dynamics entirely—its success lies precisely in the efficient construction of "semantic anchors" for the macro-environment. With minimal information redundancy, these 8 sparsely sampled frames comprehensively represent the interviewed athletes' baseline identity attributes, overall body posture inclinations, and the background atmosphere of the stadium or press conference room, thereby providing robust prior support for the downstream network to comprehend the macro-physical context in which concealed emotions transpire.

Concurrently, leveraging its massive interdisciplinary pre-trained common sense, Gemini acts as an objective psychological analysis engine. It strips away useless high-frequency noise from the visual pixel space, offering a detailed, unbiased, and logically self-consistent pure semantic parsing of the athletes' deliberately concealed behavioral details.

In contrast, the failure of the YOLO and DINO combination exposes the representation gap of current general-purpose vision foundation models in specific fine-grained micro-expression tasks. The scale of self-supervised pre-training data space for models like DINO is overly vast and complex, with optimization objectives in its latent space primarily geared toward general, open-world-level entity localization and clustering. Consequently, when presented with clean human images, the high-dimensional features extracted by DINO merely represent with extreme certainty that "the main subject in the frame is a human entity." They severely lack the feature granularity required to respond with high frequency to subtle facial muscle spasms and fleeting behavioral deviations. Without targeted micro-motion domain adaptive fine-tuning, the general dense features of foundation models are highly prone to semantic collapse in concealed action analysis, degenerating into high-dimensional redundant noise that is difficult for downstream networks to utilize.

5.2 Network Memorization Effects in Weakly Supervised Learning vs. Pseudo-optimization Traps Driven by External Verification

However, when we re-examine this "heuristic benign evolution" from a more cautious, even pessimistic dynamical perspective, we must be highly vigilant against a deeply concealed and highly probable pathological degeneration trap. When introducing vision foundation models like DINOv2/v3, which possess powerful self-supervised spatial representation capabilities, the network easily deviates from genuine cross-modal semantic alignment and instead falls into the classic shortcut learning trap of deep learning.

Experimental results indicate that while the model easily exceeds 80% accuracy on the training set, it records only about 50% performance on the test set (or submission test). This pronounced generalization gap reveals a core essence:

the high-dimensional feature manifolds extracted by DINOv2/v3 already possess astonishing instance-level discriminative power. This provides an over-parameterized network with a path of least resistance for optimization. Rather than comprehending the underlying generalization laws of data distribution, the model directly misuses visual features as high-dimensional spatial anchors. It hardens the initial pseudo-labels (and their accompanying hallucinations) into a non-parametric "implicit hash lookup table" on the training set, thereby accomplishing absolute rote memorization devoid of generalization value.

Upon this rigid, extremely overfitted ground state, other heterogeneous modalities introduced fail to provide the expected orthogonal semantic gains. Instead, due to the loss of dominant gradients, they degenerate into high-dimensional structured perturbations clinging to the decision hyperplane. This cross-modal "pseudo-adversarial interference" disrupts the original memorization homeostasis and blindly pulls at the rigid decision boundaries. As a result, the prediction distribution of the ensemble model on unseen test sets exhibits a random walk resembling Brownian motion—meaning accuracy may either increase due to accidental alignment with the test set manifold or drop due to perturbation deviations. This dilemma is further amplified when integrating external knowledge: since the label ensemble stage introduces Gemini annotations that rely entirely on external general knowledge, the associated uncertainty is directly injected into the subsequent iteration of pseudo-label learning.

At this deceptively oscillatory node, if one continues to rely absolutely on the official leaderboard—acting as an "external oracle"—for greedy selection, the entire optimization dynamics of self-training will undergo a fatal distortion. At this point, the leaderboard is essentially reduced to an implicit zeroth-order optimizer based on Monte Carlo sampling. The "high-quality pseudo-labels" we select and pass down based on performance jumps do not stem from the model ensemble reaching a superior generalization consensus. Instead, they are merely the result of a linear superposition of multi-modal random noise that happens to produce a positive error cancellation on the specific sub-sampled distribution of the current public test set.

This mechanism of inheriting labels by exploiting survivorship bias essentially accomplishes a perfect leaderboard proxy overfitting, replacing the "confirmation bias of the training set" with the "confirmation bias of the test set." Once decoupled from the specific data distribution of the current leaderboard in the future, and when faced with minor domain shifts in real-world long-tail scenarios or private test sets, this fragile sandbox built upon the coupling of strong visual memory and random modal noise will inevitably suffer a catastrophic generalization collapse.

5.3 Curse of Dimensionality in Dynamic Geometric Features vs. Generative Data Augmentation Potential

In the early stages of this study, we attempted to extract spatial offset features from 4D dynamic Gaussian point clouds based on the HRAvatar reconstruction algorithm. However, this modality was ultimately proven unsuitable as a direct discriminative input for the current classification network.

From an engineering feasibility perspective, the inverse rendering of monocular high-fidelity point clouds incurs extreme computational time overheads. Furthermore, when dealing with unpredictable, aggressive camera movements (zooming/panning) and body occlusions in wild videos, the failure rate of the underlying 3D reconstruction is exceptionally high, severely undermining the feature alignment and inference robustness of the multi-modal system in real-world scenarios.

More fatally, from an algorithmic and theoretical standpoint, the temporal offset features derived from the spatial trajectories of Gaussian point clouds construct an immensely massive, ultra-high-dimensional parameter space. In few-shot scenarios like concealed emotion understanding—where effectively labeled samples are extremely scarce—such redundant, low-level geometric features introduce massive amounts of high-frequency, unordered noise. The network encounters a classic curse of dimensionality within the highly sparse sample space, making it highly susceptible to catastrophic overfitting on specific geometric topological artifacts, which entirely masks the already faint genuine expression signals.

Although direct adoption of high-dimensional dynamic offset features might achieve statistical convergence feasibility only in ultra-large-scale datasets with massive video volumes, the failure of this exploration points to an alternative pathway with greater application value. Thinking outside the box of direct feature extraction, this high-fidelity reconstruction technology—capable of completely decoupling an individual's static physical identity from their dynamic expression offsets—exhibits immense application potential in the field of data augmentation.

Future research can leverage this rendering engine to precisely extract authentic dynamic offset parameters from a tiny fraction of long-tail categories (such as extremely rare negative concealment actions) and seamlessly retarget them across domains onto ethnic facial topologies with different feature attributes. This generative augmentation engine, based on real 3D physical deformation principles, holds the promise of synthesizing massive new samples at zero cost with unprecedented fidelity, fundamentally resolving the long-standing challenges of extreme few-shot samples and long-tail class imbalances in the field of concealed emotion from the very source of data generation.

6 Conclusion

To address the task of few-shot hidden emotion recognition, this paper proposes a compact multimodal temporal framework and redefines it as offset-aware multimodal temporal modeling. Architecturally, the framework introduces a Base-Offset cross-attention mechanism and a multi-instance learning (MIL)-based adaptive pooling layer. In terms of optimization strategy, it couples these components with a weakly supervised iterative self-training mechanism, thereby effectively capturing faint and sparse behavioral evidence in videos. In Track 3 of The 4th EI-MIGA-IJCAI Challenge, this method ultimately secured first place with a test accuracy of 0.76923.

Systematic experiments further reveal the impacts of feature engineering and learning strategies. The results demonstrate that, compared to 2D OpenPose, 3D MediaPipe—which incorporates depth information—can more accurately model spatial geometric structures and alleviate viewpoint occlusions. Meanwhile, X-CLIP video features and textual semantic priors generated by large language models (LLMs) [46] play a significant positive discriminative role within the joint network. Conversely, general-purpose vision foundation models (such as DINOv2/v3) run the risk of representation collapse in such tasks due to the lack of domain-adaptive fine-tuning tailored for micro-movements, causing their dense features to easily degenerate into high-dimensional redundant noise.

However, an in-depth learning dynamics analysis indicates that the current weakly supervised self-training mechanism possesses certain limitations. The substantial generalization gap observed in experiments between the training set accuracy (over 80%) and the test set accuracy (approximately 50%) for the DINOv2 modality suggests that over-parameterized networks are prone to shortcut learning. The model tends to convert initial pseudo-labels into implicit lookup tables on the training set, leading to severe overfitting to sample features and a failure to grasp the underlying data generalization patterns. Under this state, if the model excessively relies on public leaderboard feedback as a source for iterative selection, the optimization trajectory of self-training may deviate. In this scenario, performance gains often do not stem from a genuine generalization consensus reached by the multimodal network, but rather represent a "pseudo-generalization" phenomenon where multimodal random noise experiences positive error cancellation on the specific sampling distribution of the test set. Such a mechanism, which relies on the memorization of visual features and random noise cancellation, still carries a high risk of generalization failure when encountering domain shifts in real-world, long-tail scenarios. Consequently, when confronted with unseen sample distributions, the model's actual inference performance usually falls short of the high accuracy demonstrated on specific leaderboards.

Acknowledgments. This work was supported by Anhui Provincial Natural Science Foundation (2408085J040), National Key R&D Program of China (2024YFB3311600), Natural Science Foundation of China (62272144, 72188101), the Major Project of Anhui Provincial Science and Technology Breakthrough Program (202423k09020001), and the New Cornerstone Science Foundation through the XPLORER PRIZE.

References

1. Baltrušaitis, T., Zadeh, A., Lim, Y.C., Morency, L.P.: Openface 2.0: facial behavior analysis toolkit. In: 2018 13th IEEE International Conference on Automatic Face and Gesture Recognition, pp. 59–66 (2018)

2. Cao, Z., Hidalgo, G., Simon, T., Wei, S.E., Sheikh, Y.: OpenPose: realtime multi-person 2D pose estimation using part affinity fields. IEEE Trans. Pattern Anal. Mach. Intell. **43**(1), 172–186 (2021)
3. Carreira, J., Zisserman, A.: Quo Vadis, action recognition? a new model and the kinetics dataset. In: Proceedings of the IEEE Conference on Computer Vision and Pattern Recognition, pp. 6299–6308 (2017)
4. Chen, G., et al.: Prototype learning for micro-gesture classification. arXiv preprint arXiv:2408.03097 (2024)
5. Chen, H., Shi, H., Liu, X., Li, X., Zhao, G.: SMG: a micro-gesture dataset towards spontaneous body gestures for emotional stress state analysis. Int. J. Comput. Vision **131**, 1346–1366 (2023)
6. Cui, Y., Jia, M., Lin, T.Y., Song, Y., Belongie, S.: Class-balanced loss based on effective number of samples. In: Proceedings of the IEEE/CVF Conference on Computer Vision and Pattern Recognition, pp. 9268–9277 (2019)
7. Devlin, J., Chang, M.W., Lee, K., Toutanova, K.: BERT: pre-training of deep bidirectional transformers for language understanding. In: Proceedings of the 2019 Conference of the North American Chapter of the Association for Computational Linguistics, pp. 4171–4186 (2019)
8. Duan, H., Zhao, Y., Chen, K., Lin, D., Dai, B.: Revisiting skeleton-based action recognition. In: Proceedings of the IEEE/CVF Conference on Computer Vision and Pattern Recognition, pp. 2969–2978 (2022)
9. Team, G.: Gemini: a family of highly capable multimodal models. arXiv preprint arXiv:2312.11805 (2023)
10. Gu, J.: Performance analysis of traditional VQA models under limited computational resources. In: 2025 IEEE 6th International Conference on Pattern Recognition and Machine Learning (PRML), pp. 323–328. IEEE (2025)
11. Gu, J., et al.: Motion matters: motion-guided modulation network for skeleton-based micro-action recognition. In: Proceedings of the 33rd ACM International Conference on Multimedia, pp. 5461–5470 (2025)
12. Gu, J., Li, K., Wang, H., Akşit, K.: Text-guided fine-grained video anomaly understanding. In: Proceedings of the IEEE/CVF Conference on Computer Vision and Pattern Recognition (CVPR) Workshops, pp. 1251–1261 (2026)
13. Gu, J., Wang, F., Li, K., Wei, Y., Wu, Z., Guo, D.: MM-gesture: towards precise micro-gesture recognition through multimodal fusion. arXiv preprint arXiv:2507.08344 (2025)
14. Guo, D., Li, K., Hu, B., Zhang, Y., Wang, M.: Benchmarking micro-action recognition: dataset, methods, and applications. IEEE Trans. Circuits Syst. Video Technol. **34**(7), 6238–6252 (2024)
15. Guo, D., et al.: Mac 2024: micro-action analysis grand challenge. In: Proceedings of the 32nd ACM International Conference on Multimedia, pp. 11304–11305 (2024)
16. Huang, H., Wang, Y., Linghu, K., Xia, Z.: Multi-modal micro-gesture classification via multi-scale heterogeneous ensemble network. In: MiGA@IJCAI (2024)
17. Jiang, T., et al.: RTMPose: real-time multi-person pose estimation based on mmPose (2023)
18. Kerbl, B., Kopanas, G., Leimkühler, T., Drettakis, G.: 3D gaussian splatting for real-time radiance field rendering. ACM Trans. Graph. **42**(4), 1–14 (2023)
19. Khirodkar, R., et al.: Sapiens: foundation for human vision models. In: Leonardis, A., Ricci, E., Roth, S., Russakovsky, O., Sattler, T., Varol, G. (eds.) ECCV 2024. LNCS, pp. 206–8. Springer, Cham (2024). https://doi.org/10.1007/978-3-031-73235-5_12

20. Li, K., Gu, J., Wang, F., Wu, Z., Fan, H., Guo, D.: Ma-bench: towards fine-grained micro-action understanding. In: Proceedings of the IEEE/CVF Conference on Computer Vision and Pattern Recognition, pp. 20118–20128 (2026)
21. Li, K., et al.: Prototypical calibrating ambiguous samples for micro-action recognition. In: Proceedings of the AAAI Conference on Artificial Intelligence, vol. 39, pp. 4815–4823 (2025)
22. Li, K., Guo, D., Chen, G., Liu, F., Wang, M.: Data augmentation for human behavior analysis in multi-person conversations. In: Proceedings of the 31st ACM International Conference on Multimedia, pp. 9516–9520 (2023)
23. Li, K., Guo, D., Chen, G., Peng, X., Wang, M.: Joint skeletal and semantic embedding loss for micro-gesture classification arXiv preprint arXiv:2307.10624 (2023)
24. Li, K., et al.: Mac 2025: the 2nd micro-action analysis grand challenge. In: Proceedings of the 33rd ACM International Conference on Multimedia, pp. 14216–14221 (2025)
25. Li, K., et al.: MMAD: multi-label micro-action detection in videos. In: Proceedings of the IEEE/CVF International Conference on Computer Vision, pp. 13225–13236 (2025)
26. Li, T., Bolkart, T., Black, M.J., Li, H., Romero, J.: Learning a model of facial shape and expression from 4D scans. ACM Trans. Graph. **36**(6), 1–17 (2017)
27. Lin, H., et al.: Depth anything 3: recovering the visual space from any views. arXiv preprint arXiv:2511.10647 (2025)
28. Lin, J., Gan, C., Han, S.: TSM: temporal shift module for efficient video understanding. In: Proceedings of the IEEE/CVF International Conference on Computer Vision, pp. 7083–7093 (2019)
29. Lin, T.Y., Goyal, P., Girshick, R., He, K., Dollár, P.: Focal loss for dense object detection. In: Proceedings of the IEEE International Conference on Computer Vision, pp. 2980–2988 (2017)
30. Liu, P., Li, K., Wang, F., Wei, Y., She, J., Guo, D.: Online micro-gesture recognition using data augmentation and spatial-temporal attention. arXiv preprint arXiv:2507.09512 (2025)
31. Liu, P., et al.: Micro-gesture online recognition using learnable query points. arXiv preprint arXiv:2407.04490 (2024)
32. Liu, T., et al.: Self-supervised learning matters: a simple ensemble solution for micro-gesture recognition. arXiv preprint arXiv:2606.09261 (2026)
33. Liu, X., Shi, H., Chen, H., Yu, Z., Li, X., Zhao, G.: imigue: An identity-free video dataset for micro-gesture understanding and emotion analysis. In: Proceedings of the IEEE/CVF Conference on Computer Vision and Pattern Recognition, pp. 10631–10642 (2021)
34. Liu, Z., Ning, J., Cao, Y., Wei, Y., Zhang, Z., Lin, S., Hu, H.: Video swin transformer. In: Proceedings of the IEEE/CVF Conference on Computer Vision and Pattern Recognition, pp. 3202–3211 (2022)
35. Lugaresi, C., et al.: Mediapipe: a framework for building perception pipelines. arXiv preprint arXiv:1906.08172 (2019)
36. Ni, B., et al.: Expanding language-image pretrained models for general video recognition. In: Avidan, S., Brostow, G., Cissé, M., Farinella, G.M., Hassner, T. (eds.) ECCV 2022. LNCS, vol. 13664, pp. 1–18. Springer, Cham (2022). https://doi.org/10.1007/978-3-031-19772-7_1
37. Oquab, M., et al.: Dinov2: learning robust visual features without supervision. arXiv preprint arXiv:2304.07193 (2023)

38. Shen, X., Li, K., Wang, F., Qian, W., Jiang, J., Guo, D.: Spatial-temporal decoupled adapter for micro-gesture online recognition. arXiv preprint arXiv:2606.07355 (2026)
39. Siméoni, O., et al.: Dinov3. arXiv preprint arXiv:2508.10104 (2025)
40. Teed, Z., Deng, J.: RAFT: recurrent all-pairs field transforms for optical flow. In: Vedaldi, A., Bischof, H., Brox, T., Frahm, J.-M. (eds.) ECCV 2020. LNCS, vol. 12347, pp. 402–419. Springer, Cham (2020). https://doi.org/10.1007/978-3-030-58536-5_24
41. Vaswani, A., et al.: Attention is all you need. In: Advances in Neural Information Processing Systems (2017)
42. Wang, C., Chen, H., Wei, H., Yang, Y., Chen, Y., Zhao, G.: imigue-3k: A large-scale benchmark for micro-gesture analysis with self-supervised learning. arXiv preprint arXiv:2605.17179 (2026)
43. Wang, F., Guo, D., Li, K., Wang, M.: Eulermormer: robust Eulerian motion magnification via dynamic filtering within transformer. In: Proceedings of the AAAI Conference on Artificial Intelligence, vol. 38, pp. 5345–5353 (2024)
44. Wang, F., Guo, D., Li, K., Zhong, Z., Wang, M.: Frequency decoupling for motion magnification via multi-level isomorphic architecture. In: Proceedings of the IEEE/CVF Conference on Computer Vision and Pattern Recognition, pp. 18984–18994 (2024)
45. Wang, F., et al.: Exploiting ensemble learning for cross-view isolated sign language recognition. In: Companion Proceedings of the ACM on Web Conference 2025, pp. 2453–2457 (2025)
46. Wang, F., et al.: Xinsight: integrative stage-consistent psychological counseling support agents for digital well-being. In: Proceedings of the ACM Web Conference 2026, pp. 9297–9308 (2026)
47. Wang, M., et al.: Task-generalized adaptive cross-domain learning for multimodal image fusion. IEEE Trans. Multimedia (2026)
48. Wang, R., Li, K., Tong, A., Xu, J., Guo, D., Wang, M.: Gait emotion recognition via uncertainty-oriented class discriminative learning. IEEE Trans. Affect. Comput. 1–14 (2026)
49. Wang, W., Wang, F., Wang, Z., Shi, X., Xu, C.: Robust S3Former deep learning model for the direct diagnosis and prediction of natural organic matter (NOM) from three-dimensional excitation-emission-matrix (3D-EEM) data. Water Res. **284**, 123994 (2025)
50. Wang, Y., Liu, H., Xu, T., Shi, C., Xing, H.: Weak to strong: VLM-based pseudo-labeling as a weakly supervised training strategy in multimodal video-based hidden emotion understanding tasks. arXiv preprint arXiv:2602.08057 (2026)
51. Xu, Y., Zhang, J., Zhang, Q., Tao, D.: ViTPose: simple vision transformer baselines for human pose estimation. In: Advances in Neural Information Processing Systems (2022)
52. Yan, S., Xiong, Y., Lin, D.: Spatial temporal graph convolutional networks for skeleton-based action recognition. In: Proceedings of the AAAI Conference on Artificial Intelligence (2018)
53. Yang, Z., Zeng, A., Yuan, C., Li, Y.: Effective whole-body pose estimation with two-stages distillation. In: Proceedings of the IEEE/CVF International Conference on Computer Vision Workshops (2023)

Session 2: Explainability in Emotional Intelligence

TempA: Temporal Alignment of Human Expert-Referenced with Model-Inferred Acoustic Concepts for Speech Emotion Recognition

Seham Nasr[1(✉)], Zhao Ren[2], Sudi Murindanyi[1], and David S. Johnson[1]

[1] Center for Cognitive Interaction Technology, Bielefeld University, Bielefeld, Germany
snasr@techfak.de, smurindanyi@techfak.de, djohnson@techfak.de
[2] Cognitive Systems Lab, University of Bremen, Bremen, Germany
zren@uni-bremen.de

Abstract. Several interpretability methods for speech emotion recognition (SER) highlight salient features that drive the model's decision, but they rarely assess whether learned representations align with human expert knowledge of how emotion is expressed in speech to ensure AI decisions match human values, safety, and expectations. To address this gap, we introduce a new framework that aligns model-inferred acoustic concepts with human expert-referenced psychoacoustic concepts. Expert concepts are formalized as temporally grounded abstractions of affective vocal expression, while model-inferred concepts are extracted from internal representations using concept relevance propagation (CRP). Experiments on SER models show that the proposed temporal alignment framework, TempA, distinguishes prediction errors, augments standard performance metrics with interpretable concept-level diagnostics, and reveals limitations in existing expert acoustic–affective abstractions. TempA offers new insight into the behavior of the SER model and supports validating its learned representation by measuring its alignment with human domain expertise.

Keywords: Speech Emotion Recognition · Human-AI alignment · Explainable AI

1 Introduction

Speech Emotion Recognition (SER) has become an important component of human-centered AI systems, with applications ranging from human–computer interaction to mental health monitoring [18,33]. Recent approaches using Deep Learning (DL) models have achieved strong predictive performance [8,34]. However, their deployment in sensitive applications such as emotion recognition or

H. Chen et al. (Eds.): IJCAI 2026, CCIS 3120, pp. 125–144, 2027.
https://doi.org/10.1007/978-981-92-4529-1_10

depression detection requires more than accuracy. In particular, model decisions must be trustworthy, interpretable, reliable, and aligned with human expert knowledge to ensure their safety and reliability [30].

In response, interpretability research has produced a wide range of methods to analyze model behavior, including feature attribution, saliency visualization, and concept-based explanations [15]. Likewise, in SER, most existing approaches focus on identifying salient prosody features or time–frequency regions [19,22,25,26]. Although informative, these methods typically analyze acoustic features in isolation, offering limited validation of whether the highlighted evidence aligns with expert understanding of how emotions are expressed in speech [25]. As a result, explanations may be visually convincing while remaining weakly grounded in domain knowledge, undermining mutual understanding between humans and AI and making it harder for experts to interpret, validate, and oversee the model's decisions [30].

This limitation is compounded by the nature of speech signals, which encode emotion through complex, temporally structured acoustic patterns. Additionally, human perception of vocal emotion integrates multiple cues over extended time spans, whereas SER models may rely on short-term acoustic evidence [9]. Moreover, the evaluation of explanation quality has received limited attention, with few studies validating explanations against human judgment or domain knowledge. These factors make it challenging to assess whether SER models learn human-aligned abstractions rather than exploiting spurious or task-specific cues.

To contribute to prior work, we adopt a representational alignment perspective, providing a framework for assessing how closely the internal representations of artificial systems align with those of expert-referenced. This perspective has become increasingly important as societal expectations and regulatory pressures demand that AI systems make decisions consistent with human reasoning and values [31]. Motivated by this, we propose a computational framework to align acoustic concepts in SER, thereby improving transparency. Our approach evaluates whether model-inferred salient regions align with expert-referenced acoustic–affective concepts grounded in vocal emotion theory and psychoacoustic literature [13,14,29,36]. By explicitly comparing model-referenced concepts with expert-referenced acoustic cues, we introduce an alignment measure that quantifies the extent to which a model's prediction is explained by established vocal emotion principles, without relying on human-in-the-loop evaluation.

TempA enables representation-level error analysis in the SER model by reporting temporal alignment scores for the key speech-emotion acoustic concepts defined by human experts. It augments standard performance metrics with interpretable acoustic-level diagnostics and exposes cases where model behavior diverges from expert expectations. Grounding evaluation in formalized domain knowledge supports human trust by indicating when predictions are well justified and when uncertainty requires human review.

Our key contributions are summarized as follows:

- We introduce a computational framework, TempA, that aligns human expert-referenced with model-inferred affective concepts in SER.
- We demonstrate how acoustic concept alignment improves interpretability, error analysis, and supports trust in SER models.

2 Related Work

2.1 Explanation Methods in SER Models

Explainable AI (XAI) improves transparency and trustworthiness by helping users understand how black-box machine learning models make decisions. This is especially important in sensitive applications like healthcare and affective computing, where interpretable models support fairness, accountability, and regulatory compliance [15,23,35]. While much of this research has focused on vision and language models, interpretability in the speech domain has gained increasing attention due to the expressive and temporal nature of speech, particularly for tasks such as SER [3].

In SER, early interpretability efforts primarily focused on the model's reliance on acoustic features. For example, SHapley Additive exPlanations (SHAP) were applied to explain SER models using low-level acoustic features [26], and later extended to boosted one-dimensional acoustic representations [24]. Additionally, attention-based architectures have also been employed to implicitly learn and highlight emotionally relevant acoustic cues [19].

Several studies have explored post-hoc explanations for the operation of deep neural networks on time-frequency representations. For instance, saliency-based methods such as Grad-CAM, Local Interpretable Model-agnostic Explanations (LIME), Layer-wise Relevance Propagation (LRP), and occlusion sensitivity have been adapted to spectrogram-based SER models to visualize influential time-frequency regions [5,16]. Concept Relevance Propagation (CRP) decomposes the prediction of a black-box model by generating saliency maps through channel-concept initialization [1]. CRP explains model predictions by redistributing the output backward through the network, conditioning on concepts encoded by hidden-layer channels to yield a single attribution map. These methods highlight input regions that are important for model decisions and emphasize the importance of faithfulness in evaluating explanations [2,11]. Faithfulness measures how accurately an explanation reflects the model's true decision-making process, rather than producing visually plausible but misleading highlights [2,28]. While input perturbation sensitivity tests the reliability of saliency maps: if a saliency map highlights R_r regions as important for the model's decision, then perturbing those regions should significantly reduce the model's confidence in the selected class [23].

Other approaches have analyzed internal representations by visualizing the intersections of embeddings across emotional classes in intermediate and output layers [17]. Recently, the Beyond-Saliency framework has also enhanced explanations in SER by revealing important acoustic concepts in salient regions and correlating them with emotional arousal, thereby introducing human-level reasoning into SER explainability [22].

Despite these advances, existing SER interpretability methods largely analyze features or saliency regions but lack consistent computational mechanisms for validating model behavior against human domain knowledge [15,22]. In particular, saliency maps over spectrograms are difficult to interpret semantically, and the relevance of highlighted acoustic cues often varies across datasets and model architectures. As shown in prior work [10], such limitations can lead to undesired or misplaced user trust, especially when explanations are not grounded in domain knowledge. These challenges highlight the need for evaluative interpretability frameworks that move beyond visualization, systematically assess explanation fidelity and plausibility, and align model behavior with expert understanding of vocal emotion expression.

2.2 AI and Human Alignment Methodologies

AI–human alignment research extends interpretability by examining whether machine representations reflect human conceptual structures rather than isolated features [20,21]. Abstraction alignment formalizes this goal by comparing model-learned conceptual relationships with explicitly encoded human abstractions, measuring how well human concept hierarchies account for model uncertainty and behavior [6]. This approach shifts analysis from individual concepts to their relationships across levels of abstraction. Additionally, complementary work in computer vision has shown that deep models often fail to capture the hierarchical organization of human concepts and proposes incorporating end-user judgments to align model representations across abstraction levels, improving both interpretability and generalization [21].

Despite these advances, existing alignment methodologies remain largely focused on vision and language domains and often depend on end-user judgment or manual analysis. This leaves temporal, non-visual settings such as SER with limited tools for evaluating whether model representations align with established domain knowledge. Our work addresses this gap by introducing TempA, a fully computational, expert-grounded alignment framework for SER models. TempA operationalizes acoustic concept-level alignment in the temporal domain, enabling transparent and uncertainty-aware evaluation of model behavior. Rather than assessing models solely by task performance, TempA evaluates how consistently their internal structures and abstractions correspond to domain experts' conceptual understanding. This shift is essential for developing SER systems that are not only accurate but also trustworthy, interpretable, and suitable for human oversight.

Our alignment score is inspired by temporal interval relations [4] and the Intersection over Union (IoU) metric widely used in object detection and pattern recognition [12,27]. We quantify temporal alignment by measuring the overlap between the most salient regions identified by an XAI method and expert-referenced temporal regions associated with validated vocal emotion expression.

2.3 Non-verbal Speech Emotion Concepts

Non-verbal speech emotion concepts are prosody cues that convey emotional information independently of linguistic content and are grounded in vocal emotion expression [13,14,22,29,36]. These cues, including pitch, energy, speaking rate, jitter, shimmer, and spectral characteristics, have been extensively validated in the speech and affective computing literature as reliable indicators of emotional expression. Beyond SER, the same acoustic concepts have been successfully used in speech emotion synthesis and conversion, where controlled manipulation of prosodic and voice-quality features enables effective emotion transformation with strong perceptual validity [32]. Motivated by their theoretical grounding and empirical success, we extract expert-referenced nonverbal emotion cues from the vocal-emotion literature and use them as trusted concepts to evaluate and align SER model behavior.

3 Methodology

3.1 Problem Statement

The goal of this study is to quantify the extent to which the knowledge learned by a black-box SER model aligns with human-level, theory-driven acoustic knowledge of speech emotions. Specifically, we define this as an alignment estimation problem between model-inferred salient evidence and expert-referenced acoustic emotion concepts.

Let F denote a trained black-box SER classifier. Given an input speech clip, the audio signal is transformed into a log Mel-spectrogram representation $\mathbf{x}$, which serves as the model input. The classifier maps $\mathbf{x}$ to one of K emotion classes:

$$F : \mathbb{R}^d \to \{1, \ldots, K\}, \qquad F(\mathbf{x}) = y,$$

where y is the predicted emotion label and K denotes the number of emotion categories.

To inspect the evidence used by F, an explanation method produces a heatmap $\mathcal{S} \in \mathbb{R}^{H' \times W'}$, where H' and W' denote the frequency dimensions and the temporal of the explanation map. The most salient regions of $\mathcal{S}$ are aggregated as the model-inferred acoustic concept set $\mathcal{S}_{topk}$.

In parallel, expert-referenced acoustic concepts $\mathbf{C}_{in}$ are extracted from the same input speech signal $\mathbf{x}$. These concepts represent theory-driven acoustic emotion cues, such as temporally localized intervals or semantically meaningful acoustic patterns associated with emotional expression. The proposed framework measures the alignment between $\mathcal{S}_{topk}$ and $\mathbf{C}_i$ along two complementary dimensions: temporal alignment, which assesses whether the model matches human expert-referenced time intervals, and semantic alignment, which assesses whether the attended regions correspond to meaningful acoustic emotion concepts.

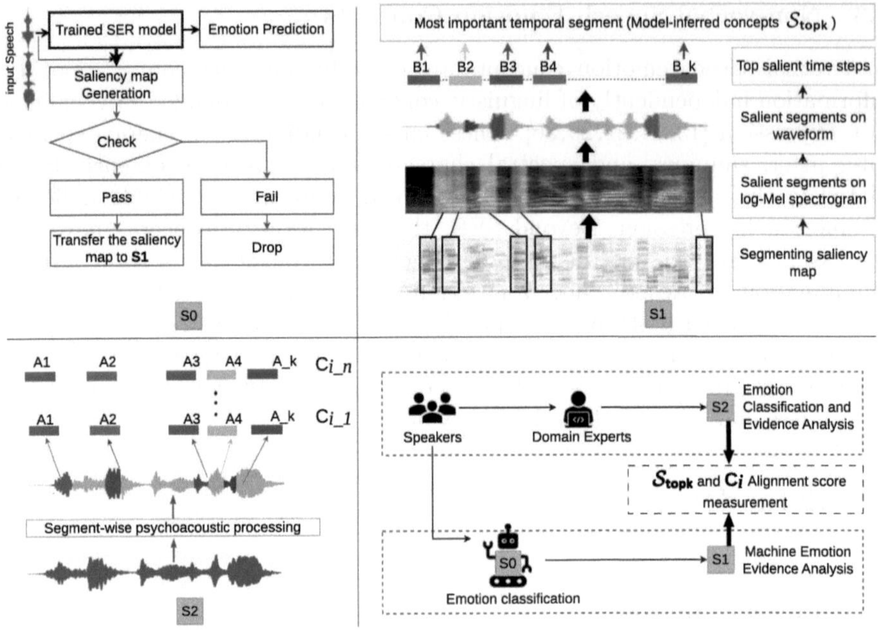

Fig. 1. Overview of the proposed TempA framework. **S0: XAI check** generates saliency maps and applies a faithfulness test, forwarding only faithful maps. **S1: Model-inferred concepts** uses *BeyondSaliency* to identify the *topk* salient temporal intervals and the human expert-referenced acoustic concepts linked to the SER model's predicted emotion. **S2: human expert-referenced acoustic concepts** analyze emotion-related theory-grounded psychoacoustic cues from the same speech input. Finally, model-inferred and expert-referenced concepts are compared using an alignment function to quantify their similarity.

Thus, as illustrated in Fig. 1 and Algorithm 1, the methodological objective is to compute an alignment score that estimates how consistently the black-box SER model bases its predictions on acoustic evidence that agrees with human-interpretable, theory-driven knowledge of speech emotion.

3.2 S0: XAI Check

We generate saliency map explanations for a trained SER model. To validate the saliency maps, we apply a top-salient region perturbation test. Given an input speech $\mathbf{x}$, SER model $F(\mathbf{x}) = y$, and a saliency map $\mathcal{S}(x)$, we perturb the top salient regions $\mathcal{S}_{topk}$ to obtain $F_{\hat{y}}(\mathbf{x}^{\mu}_{\text{top}})$. The SER model confidence drop is:

$$\Delta_{\text{top}} = F_y(\mathbf{x}) - F_{\hat{y}}(\mathbf{x}^{\mu}_{\text{top}}) \tag{1}$$

where $(\mathbf{x}^{\mu}_{\text{top}})$ is the input log Mel-spectrogram after replacing the *topk* salient windows with the mean value of the spectrogram.

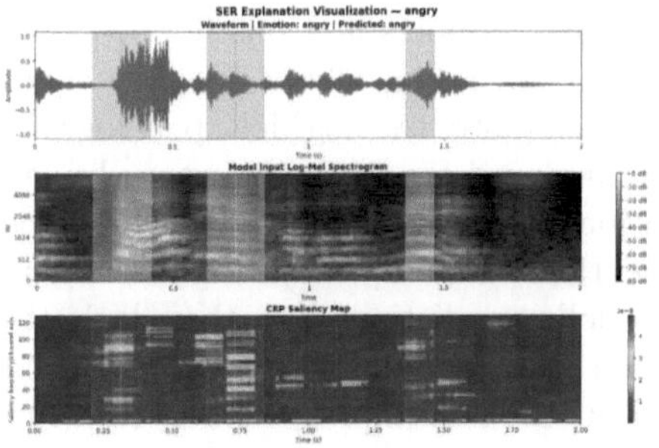

(a) saliency map

Salient regions Acoustic cues	magnitude
Loudness [sones]	02.30
Shrillness [db/kHz]	01.66
Freq. Var. [%]	71.00
Amp. Var. [dB]	00.59
Pitch lvl. [semitones re. 27.5 Hz]	22.99
Breath. [dB]	02.45

(b) Beyond Saliency

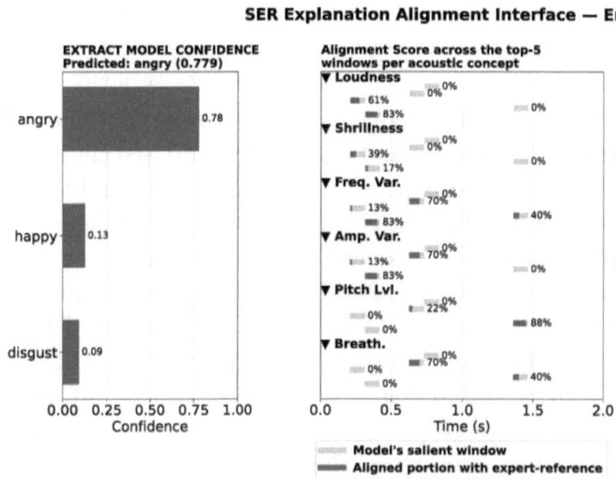

(c) Human-machine knowledge alignment

Fig. 2. Comparison of explanation outputs across methods. (a) Conventional saliency maps identify temporal regions important to the SER model's prediction but lack semantic interpretation. (b)BeyondSaliency extends (a) by linking salient regions to emotionally relevant acoustic cues. (c) Our method, TempA, further quantifies the temporal alignment between model-salient windows and expert-referenced acoustic concepts, reporting cue- and emotion-specific alignment scores.

A saliency map is faithful only if:

$$\Delta_{\text{drop}} > 0, \tag{2}$$

Only saliency maps that pass this faithfulness criterion are transferred to the next system; otherwise, unfaithful $\mathcal{S}(x)$ are discarded.

3.3 S1: Model-Inferred Acoustic Concepts

We adapt the framework in [22] to obtain a human-level interpretable SER representation. This step abstracts the reliable saliency maps into a compact set of salient temporal intervals with the most important acoustic cues, representing the model's inferred evidence for its decision, where Saliency maps are segmented, and top regions are projected onto the logMel spectrogram. Then, the resulting temporal segments are analyzed for expert-referenced vues, with"Mean" scores mapped to speech emotions. Figure 2 illustrates the expected explanation output across methods. Conventional saliency maps highlight temporal regions that influence the SER model's prediction, but they lack the semantic insights. BeyondSaliency addresses this limitation by connecting salient regions to emotion-related acoustics. Our method, TempA, extends this further by measuring the temporal similarities between model-inferred and expert-referenced acoustic concepts.

3.4 S2: Expert-Referenced Acoustic Concepts

To ground model explanations in human knowledge, we extract non-verbal acoustic cues that are well established in vocal emotion theory and psychoacoustics. These cues capture how emotions are expressed through voice quality and prosody rather than linguistic content. Specifically, we compute psychoacoustics present in Table 1. These cues have been consistently associated with emotional arousal and valence as discussed in Sect. 2.3: high-arousal emotions such as *happy*, *anger*, and *fear* are characterized by higher loudness, increased high-frequency energy perceived as shrillness, and greater pitch and voice irregularity, whereas low-arousal emotions such as *sad* and *neutral* exhibit reduced intensity, lower spectral energy, and more stable pitch patterns. The input speech signal is first segmented into fixed-duration windows after silence masking. Based on the emotion arousal category, acoustic cue values are ranked using different strategies: high-arousal emotions prioritize peak values (descending sort). In contrast, low-arousal emotions prioritize suppressed or stable regions after thresholding (ascending sort). The selected *topk* intervals C_i represent expert-referenced temporal abstractions grounded in vocal emotion theory and are later aligned with model-inferred salient regions.

Table 1. Speech emotion patterns from expert-referenced acoustic concepts [13,14,22,29].

Acoustic	Perceptual feature	High	Low
Loudness	Loudness	Happy, Anger, Fear	Neutral, Sad, Calm
High-Freq Energy	Shrillness	Happy, Anger, Fear	Neutral, Sad, Calm
Jitter	Frequency variation	Happy, Anger, Fear	Neutral, Sad, Calm
Shimmer	Amplitude variation	Happy, Anger, Fear	Neutral, Sad, Calm
F0 Mean	Pitch level	Joy, Anger	Sadness, Calm
HNR	Breathiness/hoarseness	Joy, Anger	Boredom, Sadness

Algorithm 1: Instance-wise temporal alignment between model-inferred saliency and expert-referenced concepts

Input: Speech sample $\mathbf{x}$; SER model F; target class y; explanation method $\mathcal{E}$; window width win_d; number of windows win_n; number of temporal segments k; concept set $\mathbf{C}$ (expert-referenced acoustic cues).
Output: Concept-wise alignment scores $\{\text{Alignment}(C_i)\}_{C_i \in \mathcal{C}}$.

Model-inferred concepts (salient intervals).
Generate saliency map $\mathbf{S} \leftarrow \mathcal{E}(F, \mathbf{x}, y)$
Segment $\mathbf{S}$ into win_n temporal windows of width win_d
Select *topk* salient intervals $\mathcal{S}_{topk} = \{B_j\}_{j=1}^{k}$

foreach $C_i \in C$ **do**
  Expert-referenced concepts (cue intervals).
  Segment $\mathbf{x}$ into win_n windows (same win_d) and compute cue values for C_i
  Select *topk* expert-referenced intervals $\mathbf{C}_{topk}^{(i)} = \{A_m^{(i)}\}_{m=1}^{k}$
  Temporal alignment.
  $\text{Alignment}(C_i) \leftarrow 0$
  for $j \leftarrow 1$ **to** k **do**
    $B_j = [b_j^s, b_j^e]$
    $best \leftarrow 0$
    for $m \leftarrow 1$ **to** k **do**
      $A_m^{(i)} = [a_m^s, a_m^e]$
      $ov \leftarrow \max\big(0, \min(b_j^e, a_m^e) - \max(b_j^s, a_m^s)\big)$
      $s \leftarrow \frac{ov}{b_j^e - b_j^s}$
      $best \leftarrow \max(best, s)$
    end
    $\text{Alignment}(C_i) \leftarrow \text{Alignment}(C_i) + best$
  end
  $\text{Alignment}(C_i) \leftarrow \frac{1}{k}\,\text{Alignment}(C_i)$
end
return $\{\text{Alignment}(C_i)\}_{C_i \in \mathcal{C}}$

3.5 Alignment Measurement

We quantify alignment by measuring the temporal overlap between model-inferred salient intervals and expert-referenced concept intervals. Inspired by temporal interval relations and Intersection over Union (IoU)- Sect. 2.3, we compute normalized overlap scores that capture how much of each salient region is explained by expert-defined concepts as defined in Eq. 3, and 4.

$$\text{Overlap}(B_j, A_i) = \max\big(0, \min(b_j^e, a_i^e) - \max(b_j^s, a_i^s)\big), \tag{3}$$

$$\text{Align}(B_j, A_i) = \frac{\text{Overlap}(B_j, A_i)}{b_j^e - b_j^s}, \tag{4}$$

where s and e denote the start and end of each temporal interval, respectively, $B_j = [b_j^s, b_j^e]$ denote the j-th model-inferred concept intervals, $A_i = [a_i^s, a_i^e]$

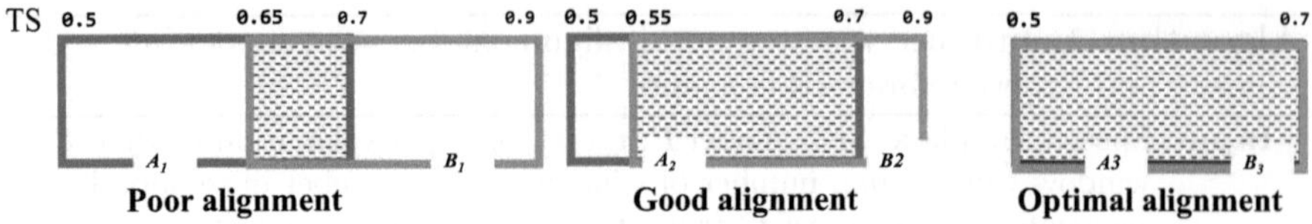

Fig. 3. Visualization of alignment calculation example between expert-referenced A and model-inferred B concept (e.g., loudness); TS is the time steps.

denote the i-th expert-referenced concept intervals. Finally, the overall alignment score for a given vocal cue is computed as the mean coverage over the *topk* salient intervals as presented in Eq. 5.

$$\text{Alignment} = \frac{1}{k}\sum_{j=1}^{k}\text{Align}(B_j). \tag{5}$$

Low or zero alignment indicates that the model relies on irrelevant regions to the expert-referenced regions, penalizing both the explanation and the underlying model behavior. Figure 3 illustrates an alignment calculation example. An optimal alignment score is observed in B_3, a poor alignment score in B_1, and B_2 is a good alignment.

3.6 Alignment Metric Validation

To verify that the proposed temporal alignment metric behaves as expected, we conducted a synthetic ground-truth validation in which the true alignment between model-inferred saliency windows and acoustic concept windows was known by construction. This validation was designed to test the correctness of the metric independently of the SER automation.

Let $B = \{B_j\}_{j=1}^{k}$ denote the set of *topk* SER models' salient temporal windows, where each window is defined as $B_j = [b_j^s, b_j^e]$. For each acoustic cue C_i, we define a synthetic ground-truth set of cue-relevant windows $A_i = \{A_m^{(i)}\}_{m=1}^{k}$, where $A_m^{(i)} = [a_m^s, a_m^e]$. Each salient window is compared against all ground-truth windows for the same cue, and the best temporal overlap is retained. The alignment score for cue C_i is then computed as in step 3.5.

We evaluated the metric under controlled synthetic conditions. In the *perfect* condition, the synthetic ground-truth windows were set to be identical to the saliency map windows, so the expected alignment score was 100%. In the *shifted* condition, each ground-truth window was shifted by half of the saliency map window length, yielding an expected overlap of 50%. In the *wrong* condition, the ground-truth windows were placed away from the saliency map windows, resulting in an expected alignment of 0%. This synthetic validation provides a sanity check for the proposed alignment framework. Since the expected alignment values are analytically known, the experiment is expected to verify that the metric returns high scores under complete temporal agreement, intermediate scores under partial agreement, and zero or near-zero scores when no temporal overlap exists.

4 Experiments

4.1 Experimental Setup

We evaluate **TempA** as a framework for measuring the temporal alignment between model-inferred acoustic concepts and human expert-referenced acoustic concepts in speech emotion recognition (SER). For each test utterance, a trained black-box SER model predicts an emotion label, and Concept Relevance Propagation (CRP) is used to obtain a post-hoc saliency map. TempA converts this saliency map into *topk* temporal windows and compares them with expert-referenced acoustic cue windows extracted from the same input speech. The goal is to determine whether the temporal evidence used by the SER model is consistent with domain knowledge about acoustic concepts such as loudness, shrillness, frequency variation, amplitude variation, pitch level, and breathiness.

4.2 Datasets

Experiments are conducted on the *CREMA-D* dataset [7], which contains 7,442 speech clips from 91 actors, including 48 male and 43 female speakers aged 20–74. The actors utter 12 sentences in six emotion categories: anger, disgust, fear, happiness, neutral, and sadness. The dataset contains approximately 5.26 hours of speech. We follow the same preprocessing setup as [22], where each audio clip is transformed into a log Mel-spectrogram before being passed to the SER model. The speech samples are resampled at 16 kHz and are preprocessed with prefix/suffix silence removal and fixed to 2 s by truncation or zero-padding. We extract log-Mel spectrograms (128 filters, 1024 FFT bins), 480-sample windows, and 240-sample hops, yielding overlapping frames with minimal loss. Datasets are split into 90% training/validation and 10% testing with balanced emotion classes.

4.3 Baselines

Since no direct prior baseline exists for the SER model-human acoustic concept alignment, we use two validation baselines. First, we perform synthetic alignment validation, where the expected temporal alignment score is known by construction. We test perfect, shifted, and wrong alignment cases, corresponding to expected scores of 100%, 50%, and 0%, respectively. Second, for XAI faithfulness, we compare the confidence drop caused by perturbing CRP-selected windows with the drop caused by perturbing random or non-salient windows of the same size.

4.4 Implementation Details

To extract model-derived knowledge, we use CRP as a post-hoc explanation method and adopt the same SER model, preprocessing setting, and XAI configuration as [22]. The CRP saliency map is segmented along the temporal axis, and the *topk* windows are selected according to cumulative saliency. We set $k = 5$ and use temporal windows of 0.1 s. The human expert-referenced acoustic cue windows are extracted using overlapping sliding windows with a 0.1 s window size and a 0.05 s hop size, resulting in 50% overlap. The CRP-derived salient windows are selected as the top-5 temporal regions after saliency aggregation and are then mapped to the corresponding time intervals in the input spectrogram.

Before alignment, we perform a perturbation-based faithfulness check. For each correctly predicted utterance, the *topk* CRP windows are perturbed in the input log Mel-spectrogram while the remaining regions are unchanged. The perturbed input is passed back to the same SER model, and the confidence drop is measured. Saliency maps are retained for TempA analysis only if the CRP-selected regions produce a meaningful confidence reduction relative to the perturbation baseline. We perturb the top-5 CRP-selected windows using mean replacement in the input log Mel-spectrogram and measure the confidence drop of the originally predicted class; a saliency map is considered faithful when this drop is positive and meaningfully larger than the drop obtained from perturbing random windows of the same size, as described in Sect. 3.2.

Expert-referenced acoustic concepts are extracted from the same utterance using acoustic features corresponding to loudness, shrillness, frequency variation, amplitude variation, pitch level, and breathiness. We extract expert-referenced acoustic cues using openSMILE eGeMAPSv02 Functionals, with 0.1 s analysis windows and a 0.05 s hop size; for high-arousal emotions (*angry*, *fear*, *happy*) we select the top-5 cue windows, while for low-arousal emotions (*sad*, *neutral*) we select the bottom-5 cue windows. All experiments are implemented in Python with PyTorch.

4.5 Evaluation Metrics

SER performance is measured using classification accuracy. TempA is evaluated using a concept-wise temporal alignment score following Sect. 3.5. The alignment score is measured using Eq. 5. The score is reported as a percentage, where 0% indicates no temporal overlap and 100% indicates full coverage of every salient window by at least one expert-referenced concept window.

Overall, the experiments evaluate TempA as an interpretable diagnostic framework that complements SER accuracy. While accuracy measures predictive correctness, TempA measures whether the temporal evidence used by the model aligns with human expert-referenced acoustic knowledge.

5 Preliminary Results

We assess the SER model's consistency with human expert knowledge using TempA, which measures temporal overlap between salient-selected model evidence and expert-referenced acoustic concept regions. This section reports how well the model's salient regions align with human-interpretable acoustic concepts.

5.1 XAI Check Results

Table 2 reports the saliency-map samples that passed/failed the faithfulness test described in Sect. 3.2. Only faithful saliency maps were retained and passed to the subsequent analysis pipeline. Among the evaluated samples, the *topk* windows for *sadness* showed the strongest faithfulness: perturbing the *topk* salient windows substantially affected the SER model, reducing its confidence from 0.9137 to 0.0435 and changing the predicted emotion to an incorrect class. For the high-arousal emotion *angry and fear*, perturbing the *topk* salient windows caused the model's confidence to drop sharply from 0.7787 to 0.0926 for angry and 0.6334 to 0.0003 for fear, changing the prediction to an incorrect class. Similarly, perturbing the salient windows for *neutral* and *happy* led to substantial confidence drops, indicating strong faithfulness of their explanations. In contrast, perturbing the top-salient windows for *disgust* produced unexpected behavior: the model's confidence increased from 0.6025 to 0.9910 instead of decreasing. This suggests that the selected salient windows did not reliably represent the prediction-driving regions. Therefore, the corresponding saliency map was considered unfaithful and excluded from further analysis.

Table 2. Faithfulness results using top-salient window perturbation, where the most salient log-mel time windows are replaced by their mean value. $\text{Pred.}_{B}/\text{Conf.}_{B}$ and $\text{Pred.}_{A}/\text{Conf.}_{A}$ denote predictions/confidences before and after perturbation, and ΔConf. is the confidence drop computed as in Sect. 3.2. The weak-faithfulness case failed.

Emotion	$\text{Pred.}_{\mathbf{B}}$	$\text{Conf.}_{\mathbf{B}}$	$\text{Pred.}_{\mathbf{A}}$	$\text{Conf.}_{\mathbf{A}}$	ΔConf.	Faithfulness
Anger	Anger	0.7787	Disgust	0.0926	**0.6861**	Strong
Fear	Fear	0.6334	Anger	0.0003	**0.6332**	Strong
Disgust	Disgust	0.6025	Disgust	0.9910	−0.3885	Weak
Happy	Happy	0.9255	Angry	0.4491	**0.4764**	Strong
Neutral	Neutral	0.5022	Happy	0.0153	**0.4870**	Strong
Sad	Sad	0.9137	Disgust	0.0435	**0.8702**	Strong

6 Alignment Scores Results

This section reports the temporal alignment between human expert-referenced acoustic concepts and SER model-inferred acoustic evidence for the faithful saliency-map instances identified in Sect. 5.1. TempA evaluates whether the model's salient windows overlap with expert-referenced cue windows for six acoustic concepts: Loudness, Shrillness, Frequency Variation (Freq. Var.), Amplitude Variation (Amp. Var.), Pitch Level (Pitch Lvl.), and Breathiness (Breath.). Alignment is reported as a percentage, where higher values indicate stronger temporal agreement between the model-inferred and expert-referenced windows. We conduct the TempA analysis at the speaker/sample level rather than averaging across utterances because acoustic emotion expression is highly speaker-dependent; averaging alignment scores can obscure individual vocal traits, speaker-specific emotion realizations, and instance-level model behavior that are central to evaluating whether a prediction is aligned with human expert knowledge.

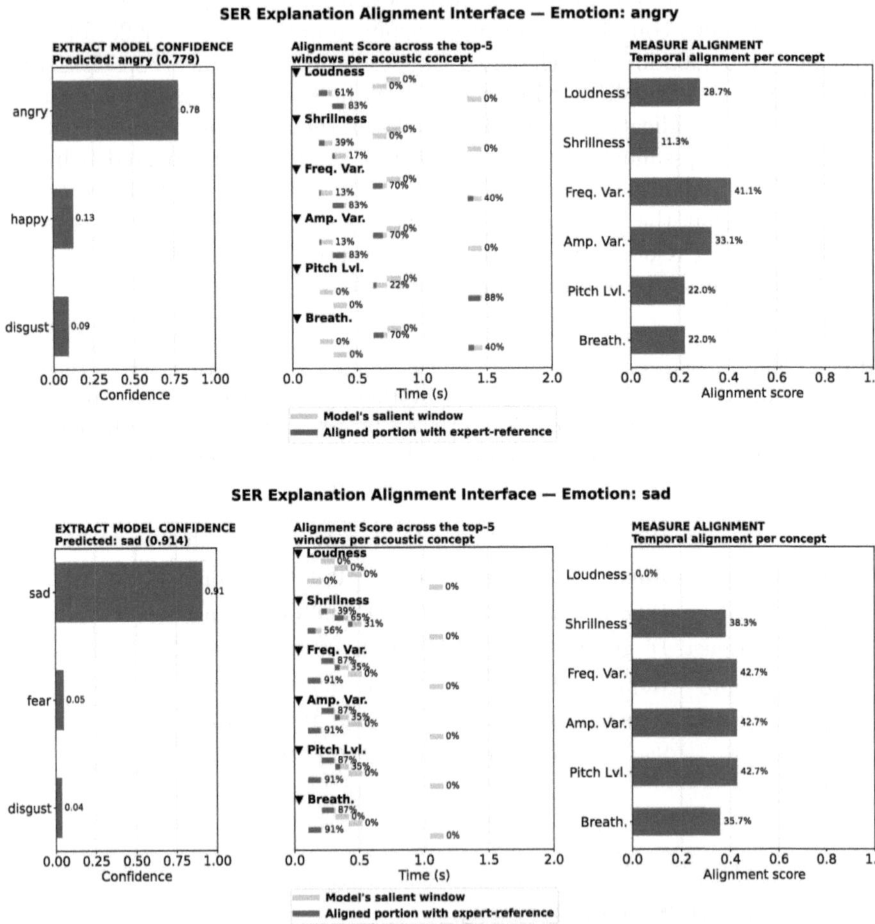

Fig. 4. Alignment Interface: SER-human alignment scores on high-arousal emotion: angry, and low-arousal emotion: sad.

Figure 4 presents the alignment results for the high-arousal emotion **angry** and the low-arousal emotion **sad**. For **angry**, the SER model predicts the target class with confidence 0.78. The strongest concept-level alignment is observed for **Freq. Var.** at 41.1%, with the highest local window-level alignment reaching 83% within the first 0.5 s. The remaining concepts show lower alignment, with **Amp. Var.** and **Loudness** providing moderate agreement, while **Pitch Lvl.**, **Breath.**, and **Shrillness** show weaker agreement. For **sad**, the model confidence is 0.91. The highest alignment scores are obtained for **Freq. Var.**, **Amp. Var.**, and **Pitch Lvl.**, each reaching 42.7%, followed by **Breath.** at 35.7%. In contrast, **Loudness** obtains 0% alignment. The window-level alignment further shows that early salient regions before 0.5 s align with several expert-referenced cues, whereas later salient windows mostly receive 0% alignment. Figure 5 shows the results for **neutral** speech, where the model confidence is 0.50. The lowest alignment is observed for **Loudness**, with a concept-level score of 7.0%, indicating limited temporal agreement with the model-inferred salient regions. In contrast, **Freq. Var.**, **Amp. Var.**, and **Pitch Lvl.** obtain the strongest alignment scores, each at approximately 41.7%. These are followed by **Shrillness** at 37.4% and **Breath.** at 34.0%. At the window level, several local matches are strong, including 91% for **Shrillness** and 83% for **Freq. Var.**, **Amp. Var.**, **Pitch Lvl.**, and **Breath.**; however, many CRP-selected windows have 0% alignment, reducing the final score averages.

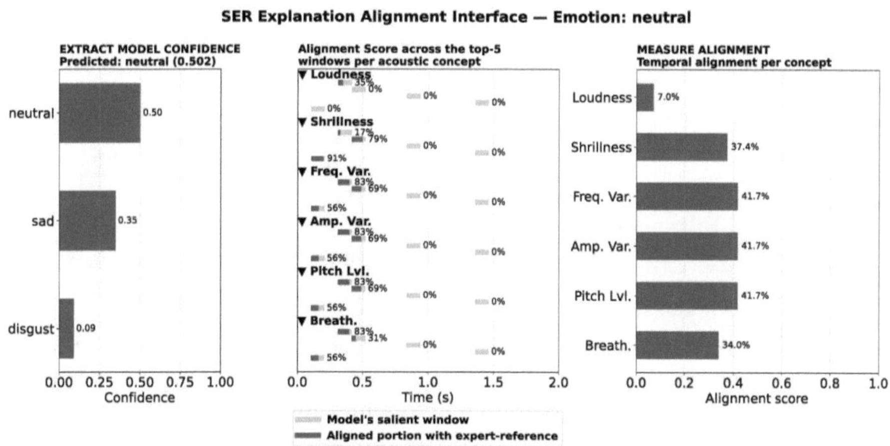

Fig. 5. Alignment Interface.

Figure 6 reports the alignment results for **happy** and **fear**. The model confidence scores are 0.66 and 0.63, respectively. For **happy**, **Freq. Var.** shows the strongest alignment at 48.0%, while **Shrillness** obtains 0%. The remaining concepts show non-zero alignment, indicating that several model-inferred salient windows overlap with expert-referenced cue regions. For **fear**, **Freq. Var.** and

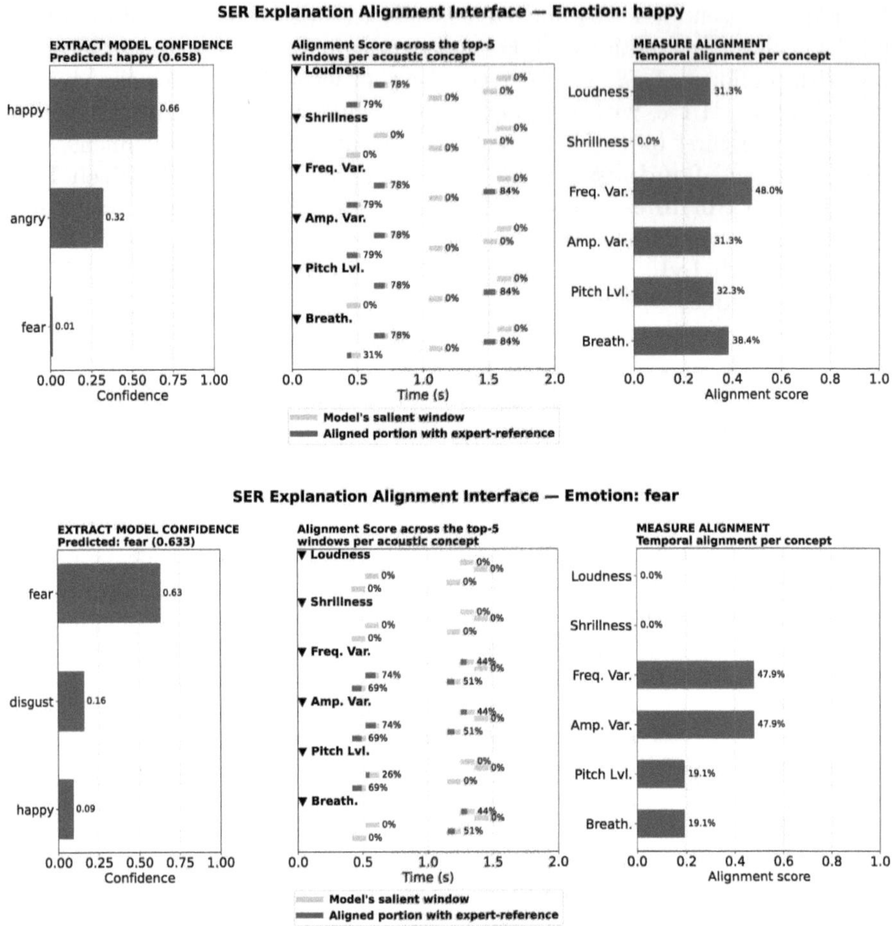

Fig. 6. Alignment Interface.

Amp. Var. both obtain the highest alignment scores at 47.9%, while **Shrillness** again shows no alignment. The local window-level panels indicate that, for both emotions, alignment is distributed across multiple temporal regions within the 2 s input rather than being confined to a single segment.

In summary, the results show that TempA provides concept-level diagnostics beyond standard SER confidence scores. Across the examined faithful explanations, **Freq. Var.** and **Amp. Var.** frequently obtain the strongest temporal alignment with model-inferred salient regions, while **Shrillness** and **Loudness** show weaker or more emotion-dependent agreement. These results indicate that TempA can quantify how closely SER model evidence corresponds to human expert-referenced acoustic concepts and can expose cases where the model's salient regions are only partially supported by the selected expert abstractions.

These findings show that TempA contributes a structured verification layer for SER explanations by linking model decisions to human expert-referenced acoustic knowledge. Instead of reporting only the predicted emotion and confidence score, TempA exposes which acoustic concepts temporally support, partially support, or fail to support the model-inferred salient regions. This interface can help domain experts inspect whether a SER decision is consistent with expected vocal evidence, identify acoustic concept-level disagreements, and distinguish reliable explanations from cases where the model relies on less interpretable or unsupported regions. Such diagnostic capability is important for developing SER systems whose decisions are required to be not only accurate but also trustworthy, interpretable, reliable, and aligned with human expert knowledge to support safer deployment.

6.1 Alignment Metric Validation Results

Independent of the SER model, we validated the temporal alignment scoring algorithm using controlled synthetic ground-truth conditions with known expected scores, as described in Sect. 3.6. The metric returned the expected outcomes for all cues: 100% alignment when ground-truth and salient windows matched, 50% alignment when windows were shifted by half the CRP-window duration, and 0% alignment when windows did not overlap. These results confirm that the metric correctly captures full, partial, and null temporal alignment without inflating scores under non-overlap.

7 Conclusion

This paper introduced **TempA**, a temporal alignment framework for comparing model-inferred acoustic concepts with human expert-referenced acoustic concepts in speech emotion recognition. Using CRP-based saliency maps from a trained black-box SER model, TempA extracts salient temporal regions and measures their overlap with theory-driven acoustic cue windows, including loudness, shrillness, frequency variation, amplitude variation, pitch level, and breathiness. Experiments on CREMA-D show that TempA complements standard classification accuracy with interpretable concept-level diagnostics, supports the analysis of SER model's predictions, and provides a structured way to examine whether model-relevant regions are consistent with human acoustic knowledge. The synthetic ground-truth validation further confirms that the proposed alignment metric behaves as expected under controlled full, partial, and non-overlapping temporal conditions.

The main contribution of this work is a reproducible framework for evaluating AI-human expert-referenced knowledge alignment in SER at the temporal and concept levels. Rather than treating saliency maps only as visual explanations, TempA operationalizes them as model-inferred acoustic concepts and compares them with expert-referenced cue intervals extracted from the same speech signal. This provides practical diagnostic value for interpreting SER models and domain knowledge insight into the extent to which learned representations reflect established acoustic-affective abstractions. TempA interface helps experts verify whether SER decisions match expected vocal evidence, identify acoustic-level disagreements, and detect unreliable explanations, supporting SER systems that are accurate, interpretable, trustworthy, and safer to deploy. However, the current evaluation is limited to CREMA-D and to a fixed set of acoustic concepts and model explanations. **Ongoing work** is extending TempA across datasets, languages, emotions, and SER architectures, including speech emotion foundation models. We will study correlations among acoustic concepts to support more comprehensive reasoning about model behavior, and we will run a human-based user study. *Code will be publicly available after finishing the ongoing work.*

References

1. Achtibat, R., et al.: From attribution maps to human-understandable explanations through concept relevance propagation. Nat. Mach. Intell. **5**(9), 1006–1019 (2023)
2. Adebayo, J., Gilmer, J., Muelly, M., Goodfellow, I., Hardt, M., Kim, B.: Sanity checks for saliency maps. In: Advances in Neural Information Processing Systems (NeurIPS), vol. 31 (2018)
3. Akman, A., Schuller, B.: Audio explainable artificial intelligence: a review. Intell. Comput. **2** (2024). https://doi.org/10.34133/icomputing.0074
4. Allen, J.F., Hayes, P.J.: Moments and points in an interval-based temporal logic. Comput. Intell. **5**(3), 225–238 (1989)
5. Becker, S., Vielhaben, J., Ackermann, M., Müller, K.R., Lapuschkin, S., Samek, W.: Audiomnist: exploring explainable artificial intelligence for audio analysis on a simple benchmark. J. Franklin Inst. **361**(1), 418–428 (2024)
6. Boggust, A., Bang, H., Strobelt, H., Satyanarayan, A.: Abstraction alignment: comparing model-learned and human-encoded conceptual relationships. In: Proceedings of the 2025 CHI Conference on Human Factors in Computing Systems, pp. 1–20 (2025)
7. Cao, H., Cooper, D.G., Keutmann, M.K., Gur, R.C., Nenkova, A., Verma, R.: Crema-d: oCrwd-sourced emotional multimodal actors dataset. IEEE Trans. Affect. Comput. **5**(4), 377–390 (2014)
8. Derington, A., Wierstorf, H., Özkil, A., Eyben, F., Burkhardt, F., Schuller, B.W.: Testing speech emotion recognition machine learning models, arXiv preprint arXiv:2312.06270 (2023)
9. Diemerling, H., Stresemann, L., Braun, T., Von Oertzen, T.: Implementing machine learning techniques for continuous emotion prediction from uniformly segmented voice recordings. Front. Psychol. **15**, 1300996 (2024)
10. Dwivedi, R., et al.: Explainable ai (xai): core ideas, techniques, and solutions. ACM Comput. Surv. **55**(9), 1–33 (2023)
11. Elisha, Y., Cohen, S., Barkan, O., Koenigstein, N.: Rethinking saliency maps: a cognitive human aligned taxonomy and evaluation framework for explanations. In: Proceedings of the AAAI Conference on Artificial Intelligence, vol. 40, pp. 3750–3758. (2026)
12. Everingham, M., Van Gool, L., Williams, C.K., Winn, J., Zisserman, A.: The pascal visual object classes (voc) challenge. Int. J. Comput. Vision **88**, 303–338 (2010)

13. Eyben, F., et al.: The GENEVA minimalistic acoustic parameter set (gemaps) for voice research and affective computing. IEEE Trans. Affect. Comput. **7**(2), 190–202 (2015)
14. Jacob, A.: Speech emotion recognition based on minimal voice quality features. In: 2016 International conference on communication and signal processing (ICCSP), pp. 0886–0890. IEEE (2016)
15. Johnson, D.S., Hakobyan, O., Paletschek, J., Drimalla, H.: Explainable ai for audio and visual affective computing: a scoping review. IEEE Trans. Affect. Comput. **16**(2), 518–536 (2024)
16. Kim, T.W., Kwak, K.C.: Speech emotion recognition using deep learning transfer models and explainable techniques. Appl. Sci. **14**(4), 1553 (2024)
17. Kumar, P., Kaushik, V., Raman, B.: Towards the explainability of multimodal speech emotion recognition. Interspeech, 1748–1752 (2021)
18. Lestari, S.W., Kahar, S., Dwi, T.: Deep learning techniques for speech emotion recognition: a review. Inter. Res. J. Sci. Technol. Educ. Manag. (IRJSTEM) **3**(2) (2023)
19. Ma, Y., Wang, W.: Msfl: explainable multitask-based shared feature learning for multilingual speech emotion recognition. Appl. Sci. **12**(24), 12805 (2022)
20. Mahner, F.P., Muttenthaler, L., Güçlü, U., Hebart, M.N.: Dimensions underlying the representational alignment of deep neural networks with humans. Nat. Mach. Intell. **7**(6), 848–859 (2025)
21. Muttenthaler, L.: Aligning machine and human visual representations across abstraction levels. Nature **647**(8089), 349–355 (2025)
22. Nasr, S., Ren, Z., Johnson, D.: Beyond saliency: enhancing explanation of speech emotion recognition with expert-referenced acoustic cues, arXiv preprint arXiv:2511.11691 (2025)
23. Nasr, S., Sikdar, S.: Indmask: inductive explanation for multivariate time series black-box models. In: European Conference on Artificial Intelligence (2024). https://api.semanticscholar.org/CorpusID:273589567
24. Nfissi, A., Bouachir, W., Bouguila, N., Mishara, B.: Unveiling hidden factors: explainable ai for feature boosting in speech emotion recognition. Appl. Intell., 1–24 (2024)
25. Norval, M., Wang, Z.: Explainable artificial intelligence techniques for speech emotion recognition: A focus on xai models. Intel. Artif. **28**(76), 85–123 (2025)
26. Pham, N.T., Nguyen, S.D., Nguyen, V.S.T., Pham, B.N.H., Dang, D.N.M.: Speech emotion recognition using overlapping sliding window and shapley additive explainable deep neural network. J. Inform. Telecommun. **7**(3), 317–335 (2023)
27. Rezatofighi, H., Tsoi, N., Gwak, J., Sadeghian, A., Reid, I., Savarese, S.: Generalized intersection over union: A metric and a loss for bounding box regression. In: Proceedings of the IEEE/CVF conference on computer vision and pattern recognition, pp. 658–666. (2019)
28. Rudin, C.: Stop explaining black box machine learning models for high stakes decisions and use interpretable models instead. Nat. Mach. Intell **1**, 206–215 (2019)
29. Scherer, K.R.: Vocal communication of emotion: a review of research paradigms. Speech Commun. **40**(1–2), 227–256 (2003)
30. Shen, H., et al.: Position: Towards bidirectional human-ai alignment. Adv. Neural. Inf. Process. Syst. 38 (2026)
31. Sucholutsky, I., et al.: Getting aligned on representational alignment, arXiv preprint arXiv:2310.13018 (2023)

32. Sun, H., Li, L., Zheng, T.F., Wang, D.: How speech is recognized to be emotional-a study based on information decomposition. In: 2021 Asia-Pacific Signal and Information Processing Association Annual Summit and Conference (APSIPA ASC), pp. 780–785. IEEE (2021)
33. Sun, Y., et al.: Weakly-supervised depression detection in speech through self-learning based label correction. IEEE Trans. Audio, Speech Lang. Process. (2025)
34. Tang, X., Huang, J., Lin, Y., Dang, T., Cheng, J.: Speech emotion recognition via cnn-transformer and multidimensional attention mechanism. Speech Commun. 103242 (2025)
35. Utomo, S., Pratap, A., Karthikeyan, P., Ayeelyan, J., Hsu, H.C., Hsiung, P.A.: When explainable artificial intelligence meets data governance: enhancing trustworthiness in multimodal gas classification. Information Fusion **125**, 103440 (2026)
36. Zhang, W., Lim, B.Y.: Towards relatable explainable ai with the perceptual process. In: Proceedings of the 2022 CHI Conference on Human Factors in Computing Systems, pp. 1–24 (2022)

HateScan: A Multimodal Web Application for Hate Speech Classification with Sentence-Level Analysis

Rathi Adarshi Rammohan(✉), Jordan Behrendt, Jost Immoor, Pablo Freitas Santos, Dennis Küster, and Tanja Schultz

Cognitive Systems Lab, University of Bremen, Bremen, Germany
{rammohan,jobe,immoor,pablo2,dennis.kuester,tanja.schultz}@uni-bremen.de

Abstract. The adverse effects of hate speech, defined as derogatory expressions that incite hatred based on immutable characteristics, have led to increasing efforts to regulate its proliferation on social media platforms. Beyond detecting hate speech, it is important to improve explainability of models to help understand the underlying mechanisms of hate speech. In this work, we present HateScan, a multimodal web application for detecting hate speech in spoken (hatespeech) and written (hatetext) content. Hand-crafted acoustic and linguistic features are used for classification, visualization, and analysis of class-wise differences. Furthermore, we propose sentence-level context modeling for hatetext detection to improve classification performance and provide deeper insights by identifying hateful segments in textual content. The application also supports sentence-level linguistic feature analysis to help examine feature variations across individual sentences. Overall, HateScan supports the understanding of individual model decisions and promotes more transparent hate speech classification.

Keywords: Hate Speech Detection · Feature Visualization · Web Application · Explainability

1 Introduction

With the rapid advancement of social media and online communication, the proliferation of harmful content has become a major challenge for digital society, with detrimental consequences [15]. One such phenomenon is hate speech, which is defined as derogatory expressions targeting individuals or groups on the basis of immutable characteristics, such as race, ethnicity, religion, or sexual orientation. Hate speech differs from abusive or offensive speech, which typically consists of insults or harsh criticism aimed at characteristics that people can change, such as political views [37]. I.e., while hate speech will typically be perceived as highly offensive, at least by members of the target group, not all offensive speech will necessarily qualify as hate speech. International regulatory bodies [11,47] have increasingly called for action against hate speech because it

H. Chen et al. (Eds.): IJCAI 2026, CCIS 3120, pp. 145–168, 2027.
https://doi.org/10.1007/978-981-92-4529-1_11

can negatively affect individuals' mental health [44,45] and, at a societal level, contribute to desensitization and discrimination [5].

As multimedia content consumption grows, hate speech increasingly manifests across modalities beyond text, including images and speech. In addition, acoustic cues such as vocal tone and speech articulation, which capture information that may be missed by textual content analysis, may further improve hate speech detection [4,20,39,40]. In this work, we use the term "hatespeech detection or classification" to refer to classification models that incorporate acoustic features, and "hatetext detection or classification" to refer to models based on linguistic features. This distinction highlights the role of different modalities in the classification process.

Hate speech detection systems are increasingly moving towards multimodal approaches, as reflected in the growing number of multimodal hate speech datasets [9,40,50,53]. Current state-of-the-art models for hatespeech and hatetext detection include self-supervised learning (SSL) models [9,50,53] and large language models (LLMs) [17,36]. However, interpreting the specific acoustic or linguistic characteristics that contribute to hate speech detection remains difficult when using the representations learned by these models [32]. Therefore, following previous work on hatespeech and hatetext detection [6,7,18,22], the present study employed hand-crafted acoustic and linguistic features.

Beyond improving hate speech classification performance as such, a better understanding of the underlying acoustic and linguistic characteristics of hate speech could help to improve our understanding of hate speech as a broader socio-emotional phenomenon. As decision-making in machine learning models is often intransparent, considerable attention has been paid to explainable hate speech classification [2,31]. Explainability aims to make model decisions understandable to humans by analysing model behaviour and interpreting the factors that influence predictions. In this work, we take a step towards explainable hate speech classification in two stages. First, we perform hatespeech and hatetext classification using hand-crafted acoustic and linguistic features. We then visualize these features to illustrate how they vary across different classes.

Second, we provide sentence-level contextual information for our hatetext models. SSL models and LLMs are known to learn contextual information relevant to hatetext detection [48]. However, the complexity of their learned representations make it difficult for interpretation of the model outcomes. In the context of hatetext classification, a lack of interpretability can lead to unexplained or unjustified content deletion [10,51]. Another challenge is the inherent non-deterministic nature of LLMs. Even with identical inputs, supposedly deterministic LLMs may produce varying outputs [3].

More traditional feature-based classification approaches often focus primarily on performance rather than interpretability [22,24]. Although some NLP approaches introduce interpretability through explainable artificial intelligence (XAI) methods, such as Local Interpretable Model-agnostic Explanations (LIME) and Shapley Additive Explanations (SHAP), these methods often focus on word-level explanations [16,31]. However, whether a sentence or text constitutes hate speech often depends on its surrounding context. What is said before or after a sentence can change its interpretation and meaning [14]. Current approaches for identifying the location of hateful content in text are often limited to keyword spotting, which evaluates the hatefulness of individual words [2,16,27,31]. This approach does not sufficiently account for context, even though it can significantly influence the meaning of sentences and texts. Moreover, a sample labeled as hateful does not necessarily mean that all parts of the sample contain hate speech [46]. We therefore propose sentence-level hatetext classification, where the model is provided with longer contextual information through neighbouring sentences. We focus the current sentence-level analysis on linguistic features because sentence boundaries are directly available from transcripts. Sentence-level acoustic feature analysis would require reliable temporal alignment between each sentence and its corresponding audio segment, which is more challenging due to pauses, overlaps and variable sentence durations in speech. Thus, in the present work, sentence-level analysis is limited to linguistic features, while acoustic features are analysed at the full-sample level.

In addition, we aim to facilitate the use of multimodal hate speech classification, feature visualization, and sentence-level explanability through an interactive web application. Existing web applications often provide only full-text classification with a single output label and limited interpretability. For example, the application by Jain and Sharma [26] takes text input and outputs a label for the full file. Similarly, the web browser plugin by Modha et al. [30] highlights entire Tweets and Facebook posts based on the predicted class. Vrysis et al. [49] present a web-based platform designed to monitor and analyse hate text against refugees and migrants in Greece, Italy, and Spain. While these applications are capable of detecting hatetext, they neither extend their functionality to audio or video inputs nor provide detailed transparency regarding the features influencing classification outcomes. The current work therefore aims to contribute an integrated web application for explainable hate speech detection and feature visualization.

The main contributions of this work are: (i) performing unimodal (speech, text) and multimodal hate speech classification using hand-crafted acoustic and linguistic features; (ii) providing sentence-level context for hatetext detection to

Table 1. List of acoustic and linguistic features.

Category	Feature Set	No. of Features
Acoustic	eGeMAPS (Extended Geneva Minimalistic Acoustic Parameter Set)	88
	MFCCs (Mel-Frequency Cepstral Coefficients)	13
	Total Acoustic Features	**101**
Linguistic	POS (Part-of-Speech)	49
	LIWC (Linguistic Inquiry and Word Count)	69
	TF-IDF (Term Frequency–Inverse Document Frequency)	50
	Total Linguistic Features	**166**

offer explanations for identified hate segments; and (iii) integrating feature visualization, model predictions, and sentence-level explanations into a web application.

2 Methodology

2.1 Data Preparation

The primary requirements for the dataset were multimodality and the availability of timestamp annotations for hate segments. Accordingly, we selected the HateMM dataset [9], which comprises 1,083 videos collected from the BitChute platform. Of these, 60.2% (652 videos) are labelled as non-hate, while 39.8% (431 videos) are labelled as hate. Additionally, the timestamps of hate segments in the hate videos are provided as arrays of tuples in the format [[start:hh:mm:ss, end:hh:mm:ss]]. In this work, videos without audio and non-English videos were excluded. After this filtering step, we retained 1,008 video files, consisting of 416 hate speech videos and 592 non-hate speech videos.

The speech data were obtained by extracting audio from the videos using the Python package `librosa` [29]. The corresponding text transcripts were generated using OpenAI's `Whisper medium` model [38]. For sentence-level contexts for hatetext classification, the hate timestamp annotations were combined with the word-level timestamp feature of the Whisper model to generate a dataset of labeled sentences. Sentences were segmented using Whisper's built-in segmentation function and labeled as hateful if at least one word in the sentence overlapped with a hate time frame. This process resulted in a sentence-level dataset comprising 10,743 hate sentences and 21,597 non-hate sentences.

2.2 Feature Extraction

A key function of the application is to extract and visualize features from speech and text data to improve the understanding and explainability of features used for hate speech classification. In the present work, we identified hand-crafted acoustic features based on vocal characteristics and linguistic features based on the structural and semantic aspects of the text.

Acoustic Features. To establish a standardized audio format, the following pre-processing steps were performed: (i) converting the audio signal to a single channel (mono) and (ii) resampling to a sampling rate of 16kHz. In this study, we selected two acoustic feature sets for analysis - the Extended Geneva Minimalistic Acoustic Parameter Set (eGeMAPS) [12] and the Mel-Frequency Cepstral Coefficients (MFCC), both extracted using the `openSMILE` toolkit [13], as shown in Table 1.

The eGeMAPs feature set provides a total of 88 features that include acoustic and paralinguistic properties, grouped into frequency-, temporal-, spectral-, and energy-related features. It comprises 25 low-level descriptors (LLDs) as acoustic parameters and their corresponding statistical derivatives (functionals) [12,43]. In addition, we extracted MFCCs, which are widely used in speech processing, affective computing, and hate speech classification tasks. MFCCs capture spectral patterns that reflect changes in the speech signal while maintaining robustness to noise [42]. Since the eGeMAPS feature set already includes the first four MFCC coefficients, we extended this representation by including the first 13 MFCC coefficients, following standard speech processing practice [19].

Linguistic Features. Similar to the audio files, the text transcripts were normalized by converting all text to lowercase and removing punctuation before feature extraction. We considered three linguistic feature sets that are useful for tasks such as sentiment analysis, hate speech detection, and psychological profiling from text data, as listed in Table 1.

- **Part-Of-Speech (POS)**: The n-gram subclass Part-Of-Speech identifies the grammatical roles of individual words in the text, providing information about the syntactic structure of sentences and the word composition [52]. The `en_core_web_sm spaCy` [1] is used to generate the POS tags and calculate their frequencies.
- **Linguistic Inquiry and Word Count (LIWC):** LIWC features are generated by categorizing words from an English dictionary into psychological and social groups, such as "anger" or "swear". These categories are used to calculate the percentage of words in each category within the text, indicating the general emotional and social tone of the text [35].
- **Term Frequency - Inverse Document Frequency (TF-IDF):** TF-IDF features denote the word relevance within the text [52]. Term frequency is calculated as the count of a specific term in a text divided by the total number of terms in that text. Inverse document frequency is calculated using the total number of samples and the number of samples containing the specific term [28]. To create a smaller TF-IDF feature set, we limited the vocabulary to the 50 most frequent terms.

2.3 Hate Speech Classification

Multimodal Classification. Unimodal (i.e., hatespeech and hatetext) and multimodal hate speech classifications were performed using a diverse set of machine learning models implemented with the `scikit-learn` library. The acoustic and linguistic features were standardized using Z-score normalization to ensure consistency across the dataset and then fed as inputs to the classifiers.

Eight classifiers were selected based on their relevance to hate speech detection and their prevalence in machine learning research, as indicated by prior work analyzing the usage frequency of classification methods between 2015 and July 2021 [52]. The chosen classifiers were Random Forest, Gradient Boosting, an ensemble model combining Random Forest and Gradient Boosting, Logistic Regression, Decision Tree, Support Vector Machine, Naive Bayes, and Multi-Layer Perceptron. To ensure reproducibility, we used a fixed random seed of 42 was applied across all models, and each model was trained independently on the training data.

The dataset was split into training and test sets using an 80:20 ratio with a random state of 42. Stratified 10-fold cross-validation was performed to maintain the class distribution across folds. Based on the 10-fold cross-validation results, we generated a classification report containing the average evaluation metrics across the entire dataset, including F1-score, recall, accuracy, area under the curve (AUC), and precision.

Multimodal hate speech classification was performed using decision-level fusion, where predictions from acoustic and linguistic models were combined for the final classification. For this purpose, we integrated the two best-performing models, one trained on acoustic features and the other on linguistic features, using soft voting to classify each audio file and its corresponding transcript as hate speech or non-hate speech [41]. Soft voting uses predicted probabilities from multiple models rather than their class labels. These probabilities were averaged, and the class with the highest average probability was selected as the final prediction.

Sentence-Level Contexts for Hatetext Classification. Beyond classifying a given sample as hate or non-hate, we aim to identify which segments of the sample contribute to the final classification decision. To this end, we used the sentence-level dataset generated from the hate timestamp annotations and performed sentence-level classification before applying full-text classification.

In the first stage, sentence-level classification was performed by incorporating neighbouring sentences as contextual input. To determine the optimal amount of context for hatetext classification, we evaluated different numbers of surrounding sentences before and after the target sentence.

Two context-selection approaches were considered. In the first approach, context was taken only from sentences preceding the target sentence. This simulates a real-time classification setting, where future content is not yet available. In the second approach, context was taken from both preceding and following sentences, allowing us to examine whether access to future content improves classification performance.

Since the number of sentences in each full text varies from a single sentence to more than 100 sentences, we adopted a flexible percentage-based context-selection strategy. As illustrated in Fig. 1, a percentage of the total number of sentences in each text was used as the context window. We explored context ranges from 0% to 20% for the before-and-after approach and from 0% to 40% for the only-before approach. These limits were selected because larger context windows showed diminishing improvements in model performance. After context windowing, linguistic feature extraction was performed. The resulting class imbalance was addressed using `RandomUnderSampler` from the `imbalanced-learn` library [23]. Machine learning models were then trained for sentence-level hate text classification.

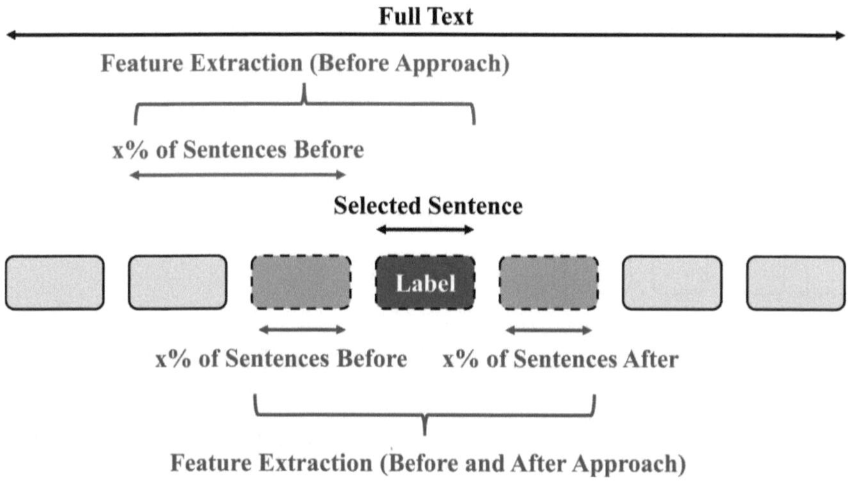

Fig. 1. Sentence-context approaches using sentences only-before (blue) and sentences before-and-after (red). (Color figure online)

In the second stage, the trained sentence-level model was used to classify each sentence of a full text to improve explainability in full-text hate classification. The sentence-level predictions were then used as additional features for predicting the label of the full text. Finally, this approach allowed us to visualize the predictions at the sentence level, which helped to make the decisions of the full-text model more transparent.

In addition to the three linguistic feature sets used for full-text classification, we introduced two sentence-level prediction-based features: hate percentage and hate sentence count. Both features were generated by applying the sentence-level model to all sentences in a text. The hate percentage represents the proportion of sentences classified as hate, while the hate sentence count represents the total number of such sentences. These two features provide complementary information. Hate percentage is more informative for longer texts, as it normalizes the amount of hateful content by text length. However, in short texts, a single misclassified sentence can substantially increase the hate percentage and lead to false positives. In such cases, hate sentence count is useful, but can be less reliable for longer texts, where the larger number of sentences increases the chance of isolated misclassifications. The influence of text length on these features is shown in Fig. 2.

The best-performing sentence-level model, selected based on the highest F1-score of the validation set, was used for feature generation in the full-text classification stage. For each full text, the sentence-level model first classified individual sentences and generated the two additional features. These features were then combined with the linguistic features extracted from the entire text and used to train the final full-text classification model.

Fig. 2. The counteracting influence of text length on the two features generated by the sentence-level model.

2.4 Model Behaviour Analysis

As a step towards making the tool's decisions more transparent, we performed additional analyses to examine the influence of linguistic features and sentence-level metrics on the hatetext model's performance.

Feature Analysis. Full-text classification relied on both linguistic features and features generated from sentence-level predictions. Therefore, understanding these features and their importance was essential for analysing the model's errors and limitations. We performed feature importance analyses for both the sentence-level and full-text classification models.

To quantify feature importance, we used Cohen's d, which indicates both the strength and direction of the difference between hate and non-hate classes. This allowed a direct comparison of the impact of individual features. We used Cohen's definitions of small, medium, and large effect sizes [8]. Cohen's d was calculated as follows [21]:

$$d = \frac{\bar{x}_1 - \bar{x}_2}{\sqrt{(s_1^2 + s_2^2)/2}}$$

Here, $\bar{x}$ denotes the group (hate and non-hate speech) mean, and s_i^2 represents the group variance. The variance was calculated by squaring the standard deviation of each group [21].

For sentence-level classification, the effect sizes of individual features were mainly used for improving explainability. To identify the strongest feature-level influences on the decision, the 20 most discriminative features were displayed together with their label direction. This showed which features were most relevant to the model's decision and whether they increased or decreased the likelihood of hate speech.

To capture broader trends, we further calculated the average and total values of hate-leaning and non-hate-leaning features. Comparing these values provided an overview of the general tendency of the sentence and helped explain how the feature patterns influenced the prediction.

Prediction Confidence and Error Analysis. To analyse the effect of confidence on prediction performance, we used the *predict_proba(X)* function from `scikit-learn` to calculate class probabilities [34]. The resulting confidence values were used to examine the relationship between prediction confidence and error rate. For distribution analysis, the predictions were grouped into 10% confidence intervals. The confidence distribution showed variations in the model's predictions. Together with the correlation between confidence and accuracy, this helped assess whether prediction confidence could indicate the likelihood of sentence-level misclassifications.

High-confidence errors were also analysed to complement the general error analysis. A confidence threshold of 0.8 was used to separate high-confidence errors from other errors. If a prediction had high confidence but was incorrect, this indicated that the sentence contained feature values strongly associated with the opposite class.

In addition, context deficiency was calculated as the difference between the number of contextual sentences expected by the model and the number actually available for a given prediction. This metric was used to analyse how missing context impacted model performance.

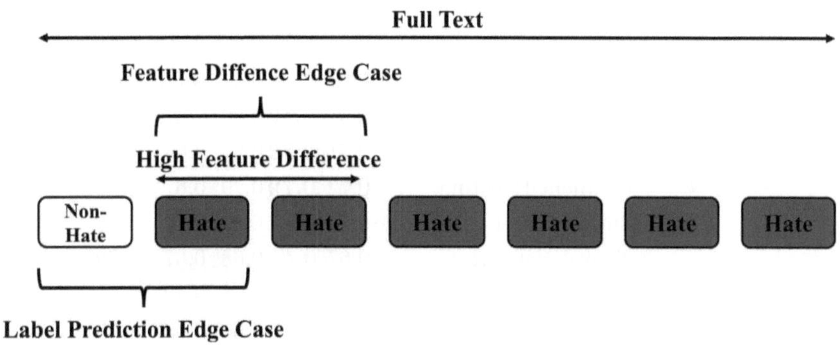

Fig. 3. Visualization of the two edge case types considered in the analysis.

Edge Case Analysis. Evaluation metrics such as accuracy, precision or recall alone were not sufficient to uncover and analyse all shortcomings of the model. To highlight more subtle limitations and issues, we examined the edge cases by analysing the differences between neighbouring sentences. This aided in identifying the situations where the distinction between hate and non-hatetext was problematic for the model.

Two categories of edge cases were considered: (i) *Feature-difference edge cases* were characterized by large feature differences in the most discriminative features between neighbouring sentences. (ii) *Label-prediction edge cases* occurred when the model's prediction changed between adjacent sentences. Together, these two groups facilitated the analysis of both data-based and model-based difficulties. They revealed how the model dealt with strong feature changes and when the model changed its predictions. Both edge case types are shown in Fig. 3. For each edge case, we performed a feature-difference analysis between the two neighbouring sentences to identify the features most likely to influence changes in the model's decision.

3 Results and Discussion

Table 2. Performance metrics across speech, text, and multimodal models. Best values within each setting are bolded, and second-best values are underlined.

Modality	Model	F1	Acc.	Prec.	Recall	AUC
Speech	Ensemble	**0.69**	**0.76**	0.74	0.66	**0.82**
	Random Forest	**0.69**	**0.76**	**0.76**	0.64	**0.82**
	Gradient Boosting	0.68	0.75	0.73	0.65	0.80
	Support Vector Machine	0.67	0.75	0.73	0.63	0.80
	Neural Network	0.67	0.73	0.67	0.67	0.76
	Logistic Regression	0.65	0.72	0.67	0.63	0.77
	Decision Tree	0.58	0.66	0.59	0.58	0.65
	Naive Bayes	0.62	0.59	0.51	**0.81**	0.70
Text	Ensemble	**0.73**	**0.79**	0.80	0.67	**0.86**
	Random Forest	0.71	**0.79**	0.81	0.64	0.85
	Gradient Boosting	0.72	**0.79**	0.79	0.67	0.85
	Support Vector Machine	0.56	0.73	**0.85**	0.42	0.82
	Neural Network	0.67	0.73	0.68	0.67	0.79
	Logistic Regression	0.54	0.71	0.78	0.42	0.79
	Decision Tree	0.65	0.71	0.66	0.63	0.69
	Naive Bayes	0.62	0.56	0.48	**0.86**	0.72
Multimodal	Multimodal Ensemble	**0.78**	**0.84**	**0.89**	**0.70**	**0.92**

3.1 Unimodal and Multimodal Hate Speech Classification

Table 2 presents the performance of the speech, text, and multimodal models on the HateMM dataset. We used the F1-score for comparison, as it is more appropriate for evaluating performance on class-imbalanced datasets.

For the acoustic features, the Random Forest and ensemble models achieved the best F1-score of 0.69. For the linguistic features, the ensemble model achieved the best performance, with an F1-score of 0.73. In both modalities, the ensemble model combined Random Forest and Gradient Boosting classifiers using soft voting. In their dataset evaluation, Das et al. [9] reported an F1-score of 0.67 using a speech model based on AudioVGG19 features extracted from raw audio waveforms. Their best-performing text model, HateXPlain, a pre-trained BERT-based model for hatetext detection, achieved an F1-score of 0.73. Our hand-crafted features achieved performance comparable to deep neural network-based approaches, demonstrating their effectiveness for hate speech classification. Furthermore, our multimodal model, which combined the best-performing speech and text models, achieved an F1-score of 0.78, exceeding the F1-score of 0.76 reported by Das et al. [9]. The best-performing speech, text and multimodal ensemble models were integrated into the web application.

3.2 Sentence-Level Hatetext Classification

Sentence-level hatetext classification was performed by varying the amount of contextual information provided to the model, i.e., the percentage of sentences only-before or before-and-after the target sentence. The best-performing ensemble text model was selected for sentence-level hatetext classification. Figure 4 shows the model performance with respect to the percentage of sentences provided as context.

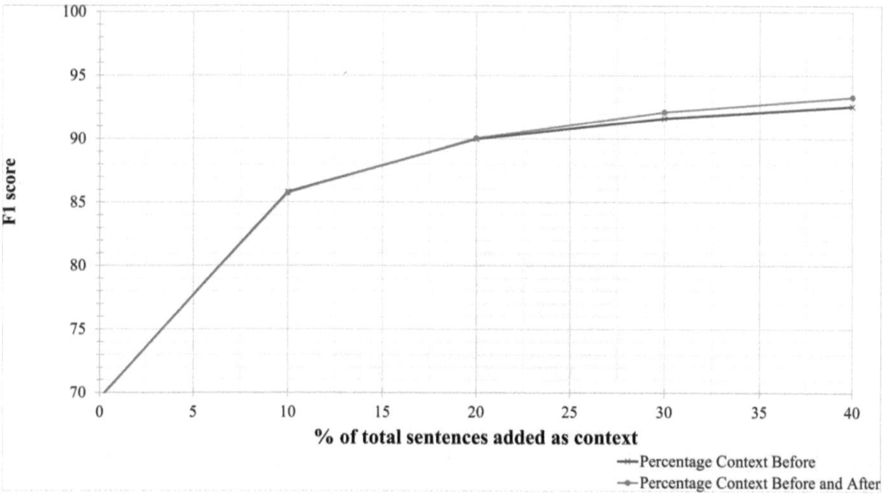

Fig. 4. Model performance according to the total percentage of sentences provided as context.

The model's performance improved as more context was provided, although the gains diminished after a certain threshold. Context expansion was stopped once the accuracy improvement fell below 1% compared to the previous step. Therefore, the maximum context window was set to 40% of the total number of sentences. Providing only-before context resulted in a slightly lower F1-score of 0.9258 compared with the before-and-after context setting, which achieved an F1-score of 0.9300. However, using only-before context would allow real-time hatetext detection, whereas subsequent context would involve more or less substantial delays. We considered this practical advantage as important enough to outweigh the marginal decrease in performance. Consequently, the 40% only-before context model was integrated into the full-text model and the web application.

Using the trained sentence-level model with 40% only-before context for full-text classification improved hatetext classification performance from an F1-score of 0.73 to 0.92, demonstrating the effectiveness of the sentence-level approach. This improvement may be attributed to the balanced dataset and the increased information given to train the sentence-level prediction model.

3.3 Model Behaviour Analysis

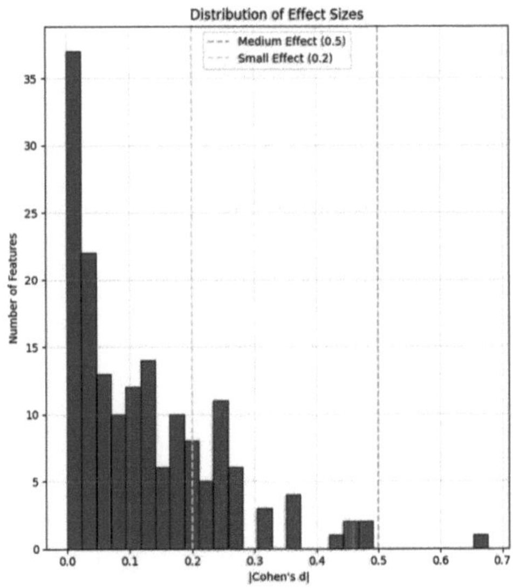

Fig. 5. Effect size distribution of linguistic features for the sentence-level model.

The analysis of feature effect sizes (Cohen's d) highlights the importance and direction of the linguistic features for hatetext classification. As shown in Fig. 5, although the sentence-level hatetext classification uses 167 features, only a few features showed meaningful individual effects. In particular, 35 features had a small effect size, and only one feature had a medium effect [8]. The remaining

131 features did not show meaningful individual effect sizes. This suggests that the model's decisions were driven by a small number of strongly discriminative features, along with a large number of other features with weaker effect size.

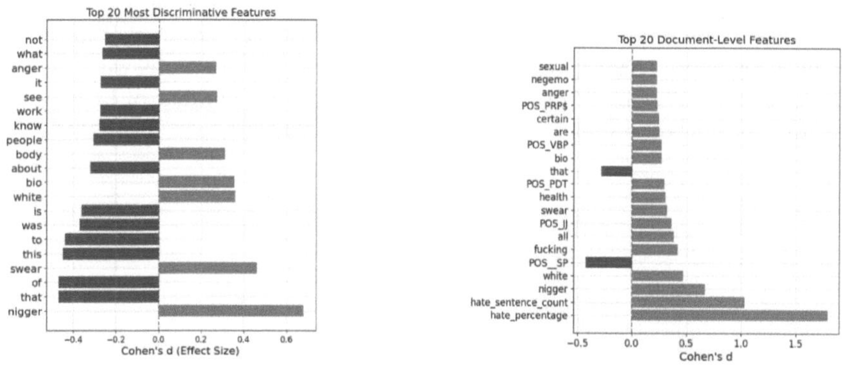

(a) Top 20 discriminative features based on effect size for sentence-level model

(b) Top 20 discriminative features based on effect size for full-text model

Fig. 6. Most discriminative linguistic features for the hatetext classification

The analysis of the 20 most discriminative features, as shown in Fig. 6a, suggests that there were more non-hate-indicating features, while hate-indicating features generally had larger effect sizes. For the HateMM dataset, the most discriminative feature was the N-gram/TF-IDF feature "nigger", which was the only feature with a medium effect size. Other hate-indicating features, such as the LIWC features "swear" and "anger", were also intuitively plausible. However, some features, such as the LIWC features "bio" and "body", and the N-gram/TF-IDF feature "white", appeared less directly associated with hate speech.

Figure 6b shows the most discriminative features for full-text classification. The inclusion of the two sentence-level prediction-based features, hate percentage and hate sentence count, had a clear effect. As expected, their direct connection to sentence-level predictions and hateful content in the text resulted in large effect sizes.

Similar to the sentence-level feature analysis, the N-gram/TF-IDF feature "nigger" remained the only original linguistic feature with a medium effect size ($|d| > 0.5$), while the remaining linguistic features showed only small effects ($|d| > 0.2$) or weaker effects. In contrast, hate percentage and hate sentence count showed large effect sizes ($|d| > 0.8$) and were hate-leaning. This indicated that the full-text model strongly relied on these sentence-level features during prediction. Interestingly, non-hate was identified more by the absence of hate-leaning features than the presence of non-hate-leaning features. Since both features were derived from the sentence-level model's predictions, the performance of the full-text model was directly linked to that of the sentence-level model.

Prediction Confidence and Error Analysis. The confidence distribution by error type, as shown in Fig. 7, indicated a clear relationship between model confidence and error rate. As confidence increased, the number of misclassifications decreased. This suggests that high-confidence predictions were more likely to be correct. We observed a disproportionately high number of false negatives. In contrast, no false positives occurred at confidence values above 90%.

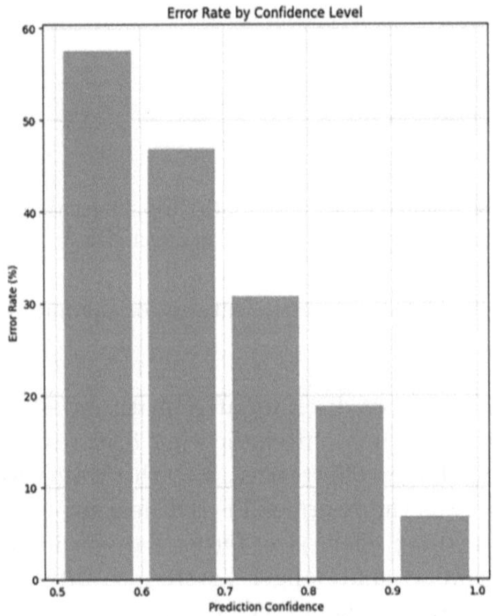

Fig. 7. The error rate for the different confidence levels.

Tables 3 and 4 summarize the feature statistics for high-confidence errors, defined as predictions with confidence above 80%. High-confidence false positives showed higher values for hate sentence count and hate percentage than true negatives, whereas high-confidence false negatives showed much lower values for these features than true positives. This suggests that these two features played a major role in high-confidence errors.

The remaining discriminative features showed different patterns across error types. In high-confidence false negatives, hate-indicating features were lower, while non-hate-indicating features were higher than in true positives. This suggested that the model failed to detect hate speech when explicit hate-related features are weak or absent.

For high-confidence false positives, however, the pattern was less consistent. Some hate-indicating features, such as "nigger" and "swear", were higher than

Table 3. High-confidence false positives compared with true negatives.

Feature	HC-FPMean	TNMean	Diff.
hate_sentence_count	26.286	0.274	26.011
hate_percentage	0.671	0.026	0.646
nigger	0.058	0.003	0.055
POS__SP	0.006	0.060	−0.054
white	0.041	0.006	0.035
POS_JJ	0.068	0.046	0.022
swear	0.036	0.017	0.020
all	0.032	0.014	0.018
POS_VBP	0.068	0.052	0.016
that	0.051	0.040	0.010

Table 4. High-confidence false negatives compared with true positives.

Feature	HC-FN Mean	TP Mean	Diff.
hate_sentence_count	0.259	28.721	−28.462
hate_percentage	0.017	0.807	−0.790
nigger	0.000	0.093	−0.093
POS__SP	0.059	0.008	0.050
swear	0.011	0.059	−0.048
fucking	0.004	0.034	−0.030
white	0.007	0.030	−0.023
POS_VBP	0.051	0.070	−0.019
bio	0.022	0.040	0−.019
all	0.011	0.029	−0.018

in true negatives, but the strongest differences came from hate sentence count and hate percentage. This suggested that high-confidence false positives were mainly driven by the sentence-level model predictions rather than by individual linguistic features alone.

Edge Case Analysis. For the edge-case analysis, we examined feature-differences and label-prediction edge cases. Since there was no clear definition of what constituted a large feature difference between neighbouring sentences, we selected the 300 sentence pairs with the highest feature differences. To avoid the influence of irrelevant features, only the 20 most discriminative features were considered.

Both edge-case types showed lower prediction confidence than the overall model average. The average confidence ranged from 58.1% to 68.3%, compared to the overall average confidence of 78.5%. This indicated that the model had greater difficulty classifying sentences in these cases. The lower accuracy further supported this observation. Feature-difference edge cases had a slightly lower accuracy of 87.5%, while label-prediction edge cases were more problematic, with an accuracy of only 61.1%.

Tables 5 and 6 summarize the most prominent feature differences for feature-difference and label-prediction edge cases, respectively. These differences helped identify which features contributed most strongly to difficult or incorrect predictions. The mean feature differences in Table 6 revealed an important characteristic of label-prediction edge cases: the features with the largest changes were not always the most discriminative features, such as "nigger" or "that", but often features with lower effect sizes, such as "bio" or "to".

Table 5. Top feature differences for feature-difference edge cases.

Feature	MeanDiff.	Max.Diff.	Sig.Cases
swear	0.0636	0.7647	159
bio	0.0478	0.3333	170
anger	0.0438	0.6667	146
nigger	0.0341	0.5173	73
is	0.0308	0.3685	114
to	0.0269	0.9157	108
of	0.0260	0.2673	111
body	0.0215	0.3333	89
that	0.0213	0.9592	78
it	0.0208	0.2764	84
this	0.0204	0.3895	85
what	0.0194	0.8889	55
was	0.0174	0.2636	54
see	0.0172	0.5000	86
know	0.0152	0.7096	46

Table 6. Top feature differences for label-prediction edge cases.

Feature	MeanDiff.	Max.Diff.	Sig.Cases
bio	0.0127	0.333	314
to	0.0108	0.171	298
swear	0.0106	0.400	217
anger	0.0100	0.400	235
of	0.0087	0.256	229
it	0.0067	0.163	148
is	0.0065	0.185	164
nigger	0.0065	0.382	76
body	0.0058	0.333	148
see	0.0057	0.182	168
that	0.0055	0.166	144
this	0.0054	0.232	126
what	0.0043	0.233	88
not	0.0041	0.395	91
was	0.0040	0.231	84

4 Web Application Integration

The hate speech classification models, feature visualizations, and sentence-level feature analyses were integrated into a web application to provide an interactive interface for researchers and content moderators. It was implemented using `Dash`, a Python framework developed by *Plotly* for building interactive web applications, particularly those requiring advanced data visualization [25].

The Hate Speech Detection Dashboard, as shown in Fig. 8 provides a responsive interface for uploading, processing, and analyzing audio and text files. The interface is organized into five main sections that guide users through the workflow:

1. **File Selection:** Users can upload files in supported formats, including *.wav*, *.mp4*, and *.txt*. If the uploaded file is a video file (*.mp4*), its audio component is extracted using *Librosa* [29]. The extracted audio, or the uploaded audio/text file, is then preprocessed and prepared for subsequent analysis.
2. **Feature Extraction:** This section enables users to extract relevant features from the uploaded files. Acoustic features are extracted from audio files, while linguistic features are extracted from text files. Once a file is selected, the *Process Acoustic Features* and *Process Linguistic Features* buttons become available. The extracted features form the basis for classification and visualization.
3. **Classification:** After feature extraction, users can initiate the classification of the uploaded files as hate speech or non-hate speech using our best-

Fig. 8. Overview of the HateScan web application

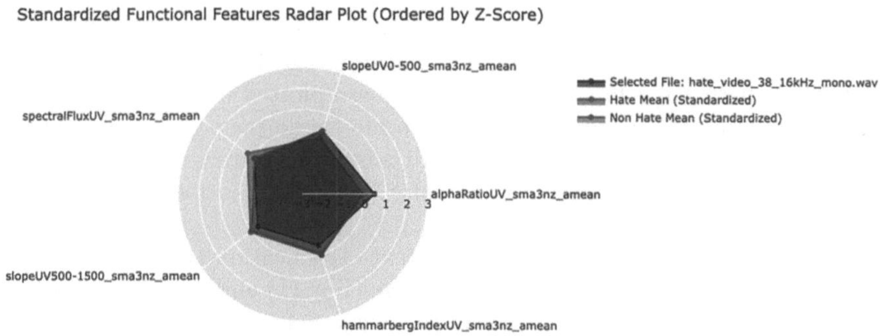

Fig. 9. An example of standardized acoustic features visualized as a radar plot

performing classifiers. The interface provides three prediction options: acoustic, linguistic, and multimodal hate speech classification. Clicking the corresponding *Predict* button initiates the selected prediction process and displays the output for the chosen file or group of files. The result is shown as either *Hate Speech Detected* in red or *No Hate Speech Detected* in green.

4. **Feature Visualization:** This section provides visual representations of the extracted features to support interpretation of the classification results. Users can select a feature set and view the extracted values as a table, radar plot, or bar plot, as seen in Fig. 9. They can also filter specific features using checkboxes in the interface.
To provide context for interpretation, the application includes reference mean values for hate speech and non-hate speech categories, computed from the HateMM dataset. These reference values are included in the table, radar plot, and bar plot visualizations.
LLDs of the acoustic features are treated separately because they provide detailed frame-level audio information and produce substantially more data

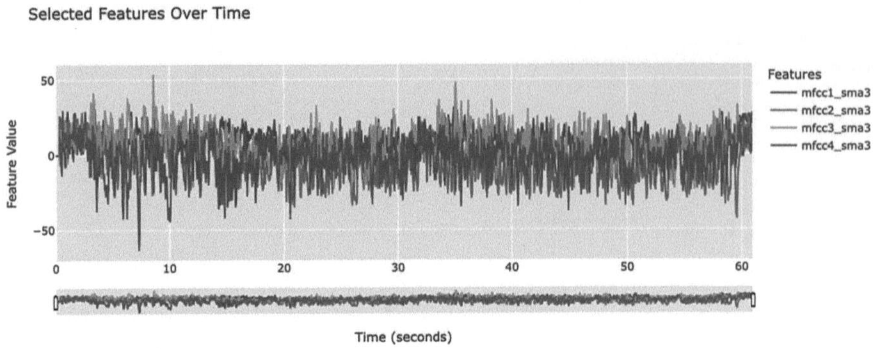

Fig. 10. Visualization of the MFCC LLDs over time

from a single file. Therefore, they are not visualized using the standard table, radar, or bar plot formats. Instead, users can select individual or all descriptors from the LLD list. An example is shown in Fig. 10, where MFCC-related LLDs, including the first four coefficients, are plotted over time. The x-axis represents the time position in the audio file, while the y-axis represents the corresponding MFCC feature values. Although this feature set is not used for prediction, it helps visualize raw acoustic feature values over time.

5. **Sentence-level Analysis:**
Sentence-level feature analyses for hatetext classification are provided in two forms: a general overview on the main page (Fig. 11) and a detailed sentence-level feature visualization on separate pages. Selecting the *Show Sentence Analysis* checkbox in the right sidebar displays the general overview generated by the sentence-level model. The overview supports the examination of the full-text model's decision by displaying the hate percentage, hate sentence count, and individual sentence predictions.
Each sentence box contains the original sentence, the predicted label, the prediction confidence, and the edge-case label, where applicable. Sentence boxes are colored green for non-hate, red for hate, and yellow for edge cases. Edge-case labels are indicated by a dashed outline in the corresponding label color. Prediction confidence is also color-coded: green for values above 0.8, yellow for values above 0.6, and red for values below 0.4, following the definition of high-confidence errors.
Clicking on a sentence opens a dedicated page with a detailed sentence prediction overview. In addition to the information shown on the main page, this view displays the context used for classification, the average and total hate- and non-hate-indicating feature values, and the most impactful individual feature values with their respective directions. This helps users to understand the model's "reasoning" behind the prediction without requiring detailed knowledge of the underlying process.
Additional information is displayed depending on the sentence position within the text. Sentences near the beginning of a text may receive less contextual

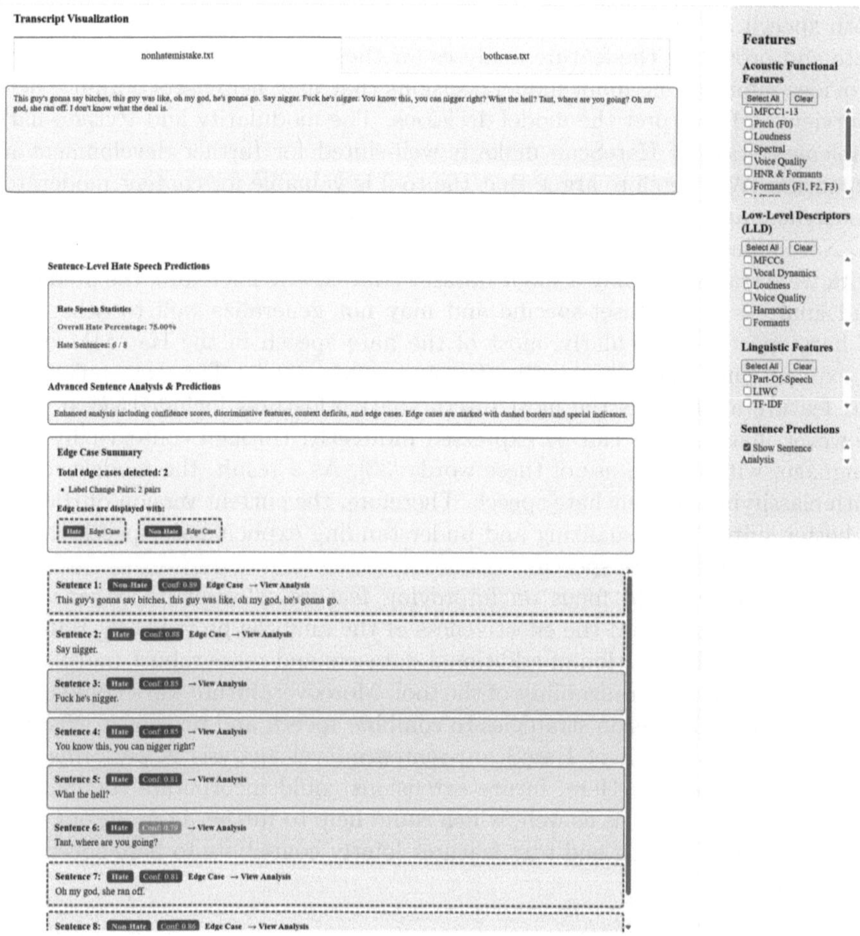

Fig. 11. General overview for the sentence-level classification visualisation.

information than intended. For example, a sentence may receive only one preceding sentence as context, even if the selected context window specifies three. This is shown as a context deficit, indicating both the expected and actual number of contextual sentences.

5 Limitations and Future Work

The development of the HateScan web application with integrated pre-trained models, feature visualization, and sentence-level feature analyses makes an important contribution to the field of hate speech detection and examines using both speech and text inputs. By highlighting which sentences are classified as hate and providing the feature analyses for these classifications, the application provides information about model decisions that may help users obtain a better overview and interpret the model decisions. The modularity and Python-native implementation of HateScan make it well-suited for further development and expansion. We therefore argue that the tool is valuable for content moderators, researchers and students who aim to analyse hate speech in multimodal data.

Neverthless, some limitations remain. The models presented in the current work were trained on only a single dataset (HateMM). Therefore, the predictions and analyses are dataset-specific and may not generalize well to other forms of hate speech. Particularly, most of the hate speech in the HateMM dataset is explicit and relies on overtly offensive swear words. This was reflected by our feature analysis as the most discriminative features included swear words. However, hate speech can be expressed indirectly, through contextually coded language, without the use of these words [33]. As a result, the models struggle with classifying implicit hate speech. Therefore, the current version of HateScan is better suited for visualizing and understanding explicit hate speech than for analysing implicit hate speech.

Future work could focus on improving feature selection to enhance both model performance and the effectiveness of the analysis provided by HateScan. The study and integration of additional datasets and more robust models could also improve the generalizability of the tool. Moreover, future work could explore better multimodal fusion strategies to combine speech and text more effectively. In the current version of HateScan, sentence-level analysis is performed only on the text modality. Here, future extensions could incorporate sentence-level acoustic feature analysis as well, which could help to further improve our understanding of how speech and text features jointly contribute to hate speech classification decisions.

6 Conclusion

Detection and mitigation of multimodal hate speech are crucial for creating safer online environments. In addition, it is important to improve our understanding of hate speech in order to facilitate a more informed discourse on how it could be better regulated. Our work takes a step in that direction by improving tool for analyzing and understanding of hate speech detection models through a web application that classifies hate speech in speech and text, visualises features, and provides sentence-level analyses to identify hate segments within a sample. Our classifiers reinforced findings from existing literature by demonstrating that multimodal classifiers effectively capture both the linguistic aspects of hate speech and their acoustic expressions. Furthermore, the performance of our sentence-level model with 40% preceding context showed its effectiveness in hatetext classification. The standard classifiers' performances metrics were supplemented by detailed feature analyses to explain the global model behaviour. This provided further insights into the factors influencing hatetext classification and complemented the local explanations offered by the application. Finally, the presented

HateScan framework is designed in a modular fashion to support flexible integration of new features, classifiers, and interpretable machine learning methods, as well as additional datasets while maintaining transparent and interpretable decision-making.

Acknowledgments. This research was funded by Deutsche Forschungsgemeinschaft (DFG; project number 465129985) and Polish National Science Centre (Narodowe Centrum Nauki, NCN; grant number 2020/39/G/HS6/00231), under the Beethoven CLASSIC 4 Polish-German funding initiative.

References

1. AI, E.: spaCy Models Usage (2023). https://spacy.io/usage/models/. Accessed 14 May 2026
2. An, J., Lee, W., Jeon, Y., Ok, J., Kim, Y., Lee, G.G.: An investigation into explainable audio hate speech detection (2024). https://arxiv.org/abs/2408.06065, https://doi.org/10.48550/arXiv.2408.06065
3. Atil, B., et al.: Non-determinism of "deterministic" LLM settings (2025). https://arxiv.org/abs/2408.04667, https://doi.org/10.48550/arXiv.2408.04667
4. Ba, Z., Yi, L., Cheng, P., Li, Q., Wang, Q., Lu, L.: Beyond content: a comprehensive speech toxicity dataset and detection framework incorporating paralinguistic cues. In: Proceedings of the AAAI Conference on Artificial Intelligence, vol. 40, pp. 21–29 (2026)
5. Bilewicz, M., Soral, W.: Hate speech epidemic. The Dynamic Effects of Derogatory Language on Intergroup Relations and Political Radicalization. Polit. Psychol. **41**(S1), 3–33 (2020). https://doi.org/10.1111/pops.12670
6. Boishakhi, F.T., Shill, P.C., Alam, M.G.R.: Multi-modal hate speech detection using machine learning. arXiv preprint arXiv:2307.11519 (2023)
7. Chebbi, S., Sekkate, S., Jebara, S.B., Adib, A.: On the use of MFCC and SWT-based features for offensive speech detection in social media. In: Proceedings of International Conference on INnovations in Intelligent SysTems and Applications (INISTA), pp. 1–6 (2021)
8. Cohen, J.: Statistical Power Analysis for the Behavioral Sciences, 2nd edn. Lawrence Erlbaum Associates (1988)
9. Das, M., Raj, R., Saha, P., Mathew, B., Gupta, M., Mukherjee, A.: HateMM: a multi-modal dataset for hate video classification. In: Proceedings of the International AAAI Conference on Web and Social Media, vol. 17, pp. 1014–1023 (2023)
10. Dobson, J.E.: On reading and interpreting black box deep neural networks. Int. J. Digit. Humanit. **5**, 431–449 (2023). https://doi.org/10.1007/s42803-023-00075-w
11. Drolsbach, C.P., Pröllochs, N.: Content moderation on social media in the EU: insights from the DSA transparency database. In: Proceedings of the ACM Web Conference 2024, pp. 939–942 (2024)
12. Eyben, F., et al.: The Geneva minimalistic acoustic parameter set (GeMAPS) for voice research and affective computing. IEEE Trans. Affect. Comput. **7**(2), 190–202 (2015)
13. Eyben, F., Wöllmer, M., Schuller, B.: openSMILE - the Munich versatile and fast open-source audio feature extractor. In: Proceedings of the ACM Multimedia (MM), pp. 1459–1462. ACM, Florence (2010)

14. Frazier, L.: On Sentence Interpretation. Springer, Dordrecht (1999). https://doi.org/10.1007/978-94-011-4599-2
15. Garcia, D., Kappas, A., Küster, D., Schweitzer, F.: The dynamics of emotions in online interaction. Roy. Soc. Open Sci. **3**(8), 160059 (2016)
16. Garg, P., Sharma, M.K., Kumar, P.: Improving hate speech classification through ensemble learning and explainable AI techniques. Arab. J. Sci. Eng. **50**, 11631–11644 (2025). https://doi.org/10.1007/s13369-024-09540-2
17. Ghorbanpour, F., Dementieva, D., Fraser, A.: Can prompting LLMs unlock hate speech detection across languages? A zero-shot and few-shot study. In: Calabrese, A., de Kock, C., Nozza, D., Plaza-del Arco, F.M., Talat, Z., Vargas, F. (eds.) Proceedings of the The 9th Workshop on Online Abuse and Harms (WOAH), pp. 413–425. Association for Computational Linguistics, Vienna (2025). https://aclanthology.org/2025.woah-1.39/
18. Gupta, V., Sharon, R., Sawhney, R., Mukherjee, D.: ADIMA: abuse detection in multilingual audio. In: Proceedings of IEEE International Conference on Acoustics, Speech and Signal Processing, pp. 6172–6176 (2022)
19. Hasan, M.R., Hasan, M.M., Hossain, M.Z.: How many mel-frequency cepstral coefficients to be utilized in speech recognition? A study with the Bengali language. J. Eng. **2021**(12), 817–827 (2021)
20. Hee, M.S., et al.: Recent advances in online hate speech moderation: multimodality and the role of large models. In: Findings of the Association for Computational Linguistics: EMNLP 2024, pp. 4407–4419 (2024)
21. Hojat, M., Xu, G.: A visitor's guide to effect sizes - statistical significance versus practical (clinical) importance of research findings. Adv. Health Sci. Educ. Theory Pract. **9**, 241–249 (2004). https://doi.org/10.1023/B:AHSE.0000038173.00909.f6
22. Ibañez, M., Sapinit, R., Reyes, L.A., Hussien, M., Imperial, J.M., Rodriguez, R.: Audio-based hate speech classification from online short-form videos. In: Proceedings of International Conference on Asian Language Processing, pp. 72–77 (2021)
23. Imbalanced-learn: Imbalanced-learn Documentation. https://imbalanced-learn.org/stable/, https://imbalanced-learn.org/stable/. Accessed 14 May 2026
24. Imbwaga, J.L., Chittaragi, N.B., Koolagudi, S.G.: Automatic hate speech detection in audio using machine learning algorithms. Int. J. Speech Technol. **27**, 447–469 (2024). https://doi.org/10.1007/s10772-024-10116-6
25. Inc., P.T.: Dash User Guide and Documentation (2024). Accessed 14 May 2026
26. Jain, A., Sharma, S.: GUI: an interface for hate speech detection using NLP technique. In: 2023 Second International Conference on Augmented Intelligence and Sustainable Systems (ICAISS), pp. 1144–1148 (2023). https://doi.org/10.1109/ICAISS58487.2023.10250527
27. Kim, J., Lee, B., Sohn, K.A.: Why is it hate speech? Masked rationale prediction for explainable hate speech detection (2022). https://arxiv.org/abs/2211.00243, https://doi.org/10.48550/arXiv.2211.00243
28. Lee, X.J., Yap, T.T.V., Ng, H., Goh, V.T.: Comparison of word embeddings for sentiment classification with preconceived subjectivity. In: International Conference on Computer, Information Technology and Intelligent Computing (CITIC 2022), pp. 488–502. Atlantis Press (2022)
29. McFee, B., et al.: librosa: audio and music signal analysis in python. SciPy **2015**(18–24), 7 (2015)
30. Modha, S., Majumder, P., Mandl, T., Mandalia, C.: Detecting and visualizing hate speech in social media: a cyber watchdog for surveillance. Expert Syst. Appl. **161**, 113725 (2020)

31. Nandini, D., Schmid, U.: Explaining hate speech classification with model agnostic methods (2023). https://arxiv.org/abs/2306.00021, https://doi.org/10.48550/arXiv.2306.00021
32. Nirmal, A., Bhattacharjee, A., Sheth, P., Liu, H.: Towards interpretable hate speech detection using large language model-extracted rationales. In: Chung, Y.L., et al. (eds.) Proceedings of the 8th Workshop on Online Abuse and Harms (WOAH 2024), pp. 223–233. Association for Computational Linguistics, Mexico City (2024). https://doi.org/10.18653/v1/2024.woah-1.17, https://aclanthology.org/2024.woah-1.17/
33. Ocampo, N.B., Sviridova, E., Cabrio, E., Villata, S.: An in-depth analysis of implicit and subtle hate speech messages. In: Proceedings of the 17th Conference of the European Chapter of the Association for Computational Linguistics, pp. 1997–2013. Association for Computational Linguistics, Dubrovnik (2023). https://doi.org/10.18653/v1/2023.eacl-main.147, https://aclanthology.org/2023.eacl-main.147/
34. Pedregosa, F., et al.: Scikit-learn: machine Learning in Python. J. Mach. Learn. Res. **12**(85), 2825–2830 (2011). http://jmlr.org/papers/v12/pedregosa11a.html, http://jmlr.org/papers/v12/pedregosa11a.html. Accessed 14 May 2026
35. Pennebaker, J.W., Francis, M.E., Booth, R.J.: Linguistic inquiry and word count: Liwc. Mahway: Lawrence Erlbaum Assoc. **71**, 2001 (2001)
36. Piot, P., Parapar, J.: Decoding hate: exploring language models' reactions to hate speech. In: Chiruzzo, L., Ritter, A., Wang, L. (eds.) Proceedings of the 2025 Conference of the Nations of the Americas Chapter of the Association for Computational Linguistics: Human Language Technologies (Volume 1: Long Papers), pp. 973–990. Association for Computational Linguistics, Albuquerque (2025). https://doi.org/10.18653/v1/2025.naacl-long.45, https://aclanthology.org/2025.naacl-long.45/
37. Puchała, D., Rammohan, R.A., Küster, D., Schultz, T., Świderska, A.: The warsaw multimodal hate speech database (WMHS): initial validation and baseline automatic detection using speech and text (2026)
38. Radford, A., Kim, J.W., Xu, T., Brockman, G., McLeavey, C., Sutskever, I.: Robust speech recognition via large-scale weak supervision. In: International Conference on Machine Learning, pp. 28492–28518. PMLR (2023)
39. Rana, A., Jha, S.: Emotion based hate speech detection using multimodal learning. arXiv preprint arXiv:2202.06218 (2022)
40. Rehman, M.Z.U., Bhatnagar, A., Kabde, O., Bansal, S., Kumar, D.N.: ImpliHate-Vid: a benchmark dataset and two-stage contrastive learning framework for implicit hate speech detection in videos. In: Proceedings of the 63rd Annual Meeting of the Association for Computational Linguistics (Volume 1: Long Papers), pp. 17209–17221. Association for Computational Linguistics, Vienna (2025). https://doi.org/10.18653/v1/2025.acl-long.842, https://aclanthology.org/2025.acl-long.842/
41. Sagi, O., Rokach, L.: Ensemble learning: a survey. WIREs Data Min. Knowl. Discov. **8**(4), e1249 (2018)
42. Schuller, B., et al.: The relevance of feature type for the automatic classification of emotional user states: low level descriptors and functionals. In: Proceedings of INTERSPEECH 2007, pp. 2253–2256. ISCA, Antwerp (2007)
43. Schuller, B.W., Steidl, S., Batliner, A., Burkhardt, F., Mazzei, L.C., Narayanan, S.: The INTERSPEECH 2013 computational paralinguistics challenge: social signals, conflict, emotion, autism. In: Proceedings of INTERSPEECH 2013. International Speech Communication Association (ISCA) (2013)

44. Soral, W., Bilewicz, M., Winiewski, M.: Exposure to hate speech increases prejudice through desensitization. Aggressive Behav. **44**(2), 136–146 (2018). https://doi.org/10.1002/ab.21737
45. Soral, W., Świderska, A., Puchała, D., Bilewicz, M.: Desensitization to hate speech: examination using heart rate measurement. Aggressive Behav. **50**(1), e22118 (2024). https://doi.org/10.1002/ab.22118
46. Swain, M., Routray, A., Kabisatpathy, P.: Databases, features and classifiers for speech emotion recognition: a review. Int. J. Speech Technol. **21**, 93–120 (2018). https://doi.org/10.1007/s10772-018-9491-z
47. United Nations: United Nations Strategy and Plan of Action on Hate Speech: Detailed Guidance on Implementation for United Nations Field Presences. Technical report, United Nations (2020). https://digitallibrary.un.org/record/3889286. Accessed 14 May 2026
48. Vaswani, A., et al: Attention is all you need. In: Advances in Neural Information Processing Systems, vol. 30. Curran Associates, Inc. (2017)
49. Vrysis, L., et al.: A web interface for analyzing hate speech. Future Internet **13**(3), 80 (2021)
50. Wang, H., Yang, T.R., Naseem, U., Lee, R.K.W.: MultiHateClip: a multilingual benchmark dataset for hateful video detection on YouTube and Bilibili. In: Proceedings of the 32nd ACM International Conference on Multimedia, MM 2024, pp. 7493–7502. Association for Computing Machinery, New York (2024). https://doi.org/10.1145/3664647.3681521
51. Wang, K., Variengien, A., Conmy, A., Shlegeris, B., Steinhardt, J.: Interpretability in the wild: a circuit for indirect object identification in GPT-2 small (2022). https://arxiv.org/abs/2211.00593
52. Zhang, W., Liu, X., Chen, L.: Exploring automatic hate speech detection on social media: a focus on content-based analysis. SAGE Open **13**(3), 21582440231181310 (2023). https://doi.org/10.1177/21582440231181311
53. Zhang, Y., et al.: DeHate: a holistic hateful video dataset for explicit and implicit hate detection. In: Proceedings of the 33rd ACM International Conference on Multimedia, MM 2025, pp. 13177–13183. Association for Computing Machinery, New York (2025). https://doi.org/10.1145/3746027.3758272

TSA-LLM: A Survey of Time-Step Alignment on Speech Large Language Models

Sheng Li[1,2(✉)], Jiyi Li[3], and Takahiro Shinozaki[1]

[1] Institute of Science Tokyo, Yokohama, Kanagawa 226-8501, Japan
sheng.li@ieee.org, shinot@ict.e.titech.ac.jp
[2] Kyoto University, Sakyo-ku, Kyoto 606-8501, Japan
[3] Hokkaido University, Sapporo, Hokkaido 060-0814, Japan
jyli@ist.hokudai.ac.jp

Abstract. Speech-language models map continuous acoustic streams to discrete language model computations, but the acoustic encoder, speech-token interface, and LLM do not share a common clock. This mismatch creates a temporal grounding problem: a model may produce a correct transcript, answer, summary, or citation while failing to indicate which audio time steps support each public output. We survey this problem as *LLM-based time-step alignment* (TSA-LLM). The survey reframes speech-text temporal alignment as an interface problem between audio time, linguistic units, and LLM steps. Using the recent literature reviewed here, we organize prior and emerging approaches into six families: monotonic ASR alignment scaffolds, forced alignment and long-form timestamping, text-aligned speech-token interfaces, LLM-based forced alignment and word timestamping, timestamp-grounded reasoning and captioning, and retrieval or contextual-biasing methods for rare entities. We then synthesize these methods into an alignment-contract view in which each grounded LLM output carries role, provenance, audio support, uncertainty, and streaming commitment status. The paper does not report a new benchmark result; instead, it provides a survey taxonomy, a design vocabulary, and an evaluation checklist for future work on trustworthy speech LLMs.

Keywords: Speech large language models · Time-step alignment · Temporal grounding · Forced alignment · Speech-token alignment · Timestamped reasoning · Survey

1 Introduction

Speech large language models (speech LLMs) increasingly support tasks that go beyond automatic speech recognition: spoken question answering, audio-grounded summarization, meeting analysis, speech-to-speech translation, speaker-attributed reasoning, retrieval-augmented transcription, and conversational assistants. These systems typically combine a frame-level speech encoder,

H. Chen et al. (Eds.): IJCAI 2026, CCIS 3120, pp. 169–184, 2027.
https://doi.org/10.1007/978-981-92-4529-1_12

an adapter or tokenizer that reduces the audio sequence, and an LLM that performs language generation or reasoning. The architecture is powerful, but it hides an important question: which part of the audio supports each generated word, timestamp, answer, citation, or reasoning claim?

This paper surveys that question as TSA-LLM, short for LLM-based time-step alignment. We use time-step alignment to mean the assignment of LLM conditioning or generation steps to explicit supports in audio time. The support can be a word boundary, phone span, speaker turn, prosodic cue, non-speech event, retrieved entity occurrence, or a deliberately ungrounded role, such as formatting or instruction-following. The target is broader than conventional word timestamping. A transcript word needs a time span, but an answer sentence may need a set of evidence spans; a speaker attribution needs both lexical evidence and diarization provenance; and a streaming assistant needs to state whether a citation is provisional, committed, or revised.

The central difficulty is a clock mismatch. Audio has a sample or frame clock; recognition modules often use phone, grapheme, subword, word, or segment clocks; and LLMs operate on prompt tokens, hidden states, evidence tokens, generated tokens, and tool-call steps. A speech LLM can collapse these clocks and still score well on word error rate or answer accuracy. It may rely on parametric knowledge rather than spoken evidence, cite a plausible but incorrect timestamp, or summarize long-form speech while losing synchronization. Standard text metrics do not expose these failures.

Existing research contains many ingredients for solving TSA-LLM. Classical methods such as CTC, recurrent neural network transducers, attention-based sequence-to-sequence models, monotonic attention, and continuous integrate-and-fire (CIF) provide monotonic alignment scaffolds [7,10,16,17,39]. Self-supervised encoders and strong acoustic backbones provide representations from which speech-token interfaces are built [1,19,23]. Forced alignment and long-form timestamping locate transcript words or segments in audio [2,31,38]. More recent work explicitly connects alignment to speech LLMs: text-aligned speech tokenization, LLM-based forced alignment, word-level timestamp prediction, timestamp-grounded reasoning, speaker-attributed ASR, timestamped captioning, retrieval-augmented speech LLMs, and modality-reasoning alignment [13,24–26,29,33,45–47].

Our contribution is a survey and synthesis rather than a new architecture. First, we formalize the TSA-LLM alignment object and distinguish acoustic, linguistic, and LLM clocks. Second, we organize current methods from the reviewed literature into a taxonomy that shows what each family contributes and where it falls short. Third, we argue for an *alignment contract*: a public interface in which speech LLMs expose typed, uncertainty-aware audio supports for grounded outputs. Finally, we propose an evaluation checklist that separates boundary precision, grounded faithfulness, calibration, long-form drift, streaming latency, and provenance accuracy.

2 Problem Definition: LLM Steps Need Audio-Time Supports

Let $x_{1:T}$ denote acoustic frames or low-level speech features, $h_{1:M} = E_\phi(x_{1:T})$ denote encoder states, $v_{1:N} = A_\psi(h_{1:M})$ denote speech or evidence tokens supplied to an LLM, and $y_{1:K}$ denote generated output tokens. A conventional recognizer estimates a transcript, sometimes with word timestamps. A TSA-LLM system estimates a support structure for LLM-visible steps. For a step k, the target alignment object can be written as

$$a_k = \big(r_k, q_k(t), S_k, c_k, p_k, m_k\big), \tag{1}$$

where r_k is a role label, $q_k(t)$ is a distribution over audio time or encoder indices, $S_k = [s_k, e_k]$ is an interval or set of intervals when an interval summary is meaningful, c_k is a confidence or calibrated uncertainty estimate, p_k records provenance such as audio, retrieval, instruction, or parametric prior, and m_k records streaming commitment status.

This object may be attached at different granularities. In ASR, it may attach to words or phones in the transcript. In speech question answering, it may attach to answer claims rather than to individual answer tokens. In diarization-aware reasoning, it may attach to a tuple (claim, time span, speaker). In captioning, it may attach to event spans. In retrieval-augmented speech understanding, it may attach to both heard evidence and retrieved context. The general problem is therefore not simply "predict timestamps", but "make LLM computation auditable against audio time".

2.1 Three Clocks

The acoustic clock indexes samples, frames, encoder states, and local acoustic events. The linguistic clock indexes phones, graphemes, subwords, words, speaker turns, discourse segments, and captions. The LLM clock indexes prompt tokens, speech/evidence tokens, hidden states, generated tokens, citations, and tool calls. Many failures arise because a model exposes only one clock. For example, a word-level aligner exposes a linguistic-to-audio relation, but it does not state which transcript words support a generated answer. Conversely, a speech LLM may expose a generated citation string, but without a calibrated relation back to encoder time.

2.2 Alignment Objects and Roles

A useful survey distinction is the difference between an *alignment object* and an *alignment mechanism*. CTC paths, transducer lattices, monotonic attention weights, CIF firing weights, adapter pooling maps, timestamp tokens, and span heads are mechanisms. The object needed by downstream users is a typed support relation. Role labels are essential: not every LLM token should be forced to have an audio span. A grounded lexical token, a speaker tag, a prosodic judgment, a retrieved entity, a formatting token, and an ungrounded reasoning bridge have different provenance requirements.

Table 1. Survey taxonomy for LLM-based time-step alignment.

Family	Representative methods	TSA-LLM question
Monotonic ASR scaffolds	CTC, RNN-T, monotonic attention, CIF, transducer alignment regularization	How can frame-level alignment uncertainty be preserved for LLM reasoning instead of collapsed into text?
Forced alignment and timestamping	MFA, Whisper-style timestamps, WhisperX, internal Whisper aligners, multilingual forced aligners	How can transcript boundaries be lifted into evidence supports for LLM claims?
Speech-token and adapter interfaces	self-supervised units, PAST, TASTE, TASTE-Streaming, Latent Speech-Text Transformer, TADA	How can audio be compressed to LLM length while retaining token-to-time provenance?
LLM-based aligners and timestamp heads	LLM-ForcedAligner, In-Sync, timestamp-aware ASR heads	How can LLM semantics help alignment without hallucinating plausible times?
Reasoning, captioning, and attribution	Listen First, Speaker-Reasoner, TAC, CASTELLA, ALARM, TARS	Which answer, caption, or speaker-attribution claim is licensed by which audio span?
Retrieval and contextual biasing	kNN-CTC, DYNAC, SEAL, CLAR	Did a rare entity come from audio evidence, retrieval context, or both?

2.3 Desiderata

A TSA-LLM alignment should satisfy five properties. *Coverage* means that salient audio evidence is attached to relevant outputs. *Locality* means that a grounded output is supported by a compact or interpretable region, unless the task genuinely requires distributed evidence. *Order sensitivity* means that transcript-like outputs respect speech order, while summaries or answers may aggregate and reorder evidence. *Calibration* means that confidence scores correspond to empirical timing and provenance error. *Faithfulness* means that cited spans matter: removing or perturbing the cited audio should affect the probability or validity of the claim it supports.

3 Survey Taxonomy

Table 1 summarizes the major method families. The taxonomy is intentionally interface-oriented. The question is not whether a method is "an aligner" in a narrow sense, but whether it contributes an auditable relation between audio time and an LLM-visible computation step.

3.1 Axis 1: Granularity

Granularity ranges from frames and phones to subwords, words, speaker turns, acoustic events, captions, and answer claims. Frame-level alignment is needed for precise boundaries, but LLM applications frequently need claim-level grounding. A meeting-summary sentence may compress several turns; a spoken QA answer may cite a short phrase; a speaker-aware answer may require the intersection of text evidence and diarization evidence. Surveying TSA-LLM therefore requires moving across granularities rather than privileging a single one.

3.2 Axis 2: Direction of Alignment

Alignment may be audio-to-text, text-to-audio, or LLM-step-to-audio. Audio-to-text alignment asks where labels occur in an acoustic stream. Text-to-audio forced alignment assumes a transcript and estimates boundaries. LLM-step-to-audio alignment is broader: it asks which audio evidence supports generated or latent LLM steps, including steps that do not correspond to transcript words. This third direction is the defining direction for TSA-LLM.

3.3 Axis 3: Inference Timing

Offline systems can use full future context. Streaming systems must decide when enough evidence has arrived. A timestamp produced after a full recording is not equivalent to a timestamp attached to a live partial answer. Streaming tokenization, delay-penalized objectives, chunkwise models, and trans-chunk architectures make this distinction explicit [42,46,48]. For speech LLMs, the relevant latency is not just token emission time; it is the time at which an output can be supported by a reliable citation.

4 Classical Alignment Scaffolds Revisited

Classical ASR alignment mechanisms remain central to TSA-LLM. They are not obsolete preprocessing; they are sources of monotonic constraints, path uncertainty, and weak supervision for LLM-level grounding.

4.1 CTC and Transducer Scaffolds

CTC marginalizes over monotonic frame-label paths and can be trained from transcripts without frame labels [17]. For an utterance label sequence $u_{1:L}$, a CTC-style scaffold defines path probabilities over frame labels:

$$p_\phi(\pi \mid x) = \prod_{m=1}^{M} p_\phi(\pi_m \mid h_m), \qquad B(\pi) = u_{1:L}, \tag{2}$$

where B removes blanks and repeated labels. The Viterbi path gives one alignment, but the posterior over paths provides uncertainty. In TSA-LLM, that uncertainty can guide evidence token construction or constrain timestamp decoders.

Recent CTC work targets issues directly relevant to LLM grounding. Delay-penalized CTC improves emission timing for streaming [48]. Adaptive maximum-entropy CTC, blank-aware knowledge distillation, and graph CTC address blank behavior, distillation, and pronunciation alternatives [12,15,22]. Robust and multilingual variants address group disparities, accents, and language coverage [3,28,54]. Pseudo-labeling and simulation approaches use CTC structure to support adaptation when dense time labels are unavailable [20,36,37]. These methods are not always presented as LLM interfaces, but they provide the frame-to-symbol scaffolds that LLM-level alignment needs.

Transducers provide another monotonic family by combining acoustic context with label history [16]. They are attractive for streaming and endpoint-sensitive decisions, but their internal alignments must be surfaced if they are to support LLM outputs. Recent work on transducer alignment regularization and segmented augmentation illustrates how alignment constraints can shape transducer behavior [27,32]. For TSA-LLM, a transducer lattice can be used as a candidate-span generator rather than only as a decoder.

4.2 Attention, Monotonic Attention, and CIF

Attention-based encoder-decoder ASR learns source-target dependencies [7]. Raw attention weights, however, are not automatically faithful explanations: they may be diffuse, layer-specific, prompt-sensitive, or affected by decoding heuristics. Monotonic attention enforces order and provides a more alignment-like interface [39]. CIF accumulates acoustic evidence until firing events produce token-level representations [10]. These mechanisms are useful for TSA-LLM only if their intermediate mappings are preserved. When an adapter compresses thousands of frames into a small number of speech tokens, it must expose how each token maps back to time.

4.3 Forced Alignment and Long-Form Timestamping

Forced alignment maps known transcripts to audio spans and is widely used in corpus preparation, phonetics, pronunciation analysis, and speech evaluation. The Montreal Forced Aligner is a representative practical workflow [31]. Whisper and WhisperX made long-form weakly supervised transcription and time-accurate timestamping prominent in modern pipelines [2,38]. Recent work further studies Whisper's internal word alignment behavior and real-time multilingual forced alignment [40,50]. A forced-alignment benchmark provides a useful evaluation anchor [9].

For TSA-LLM, forced alignment is necessary but insufficient. It can locate transcript words, but it does not tell us which words support a later answer, summary sentence, speaker claim, or retrieval decision. The survey implication is that transcript timestamps should be treated as an intermediate layer in a larger evidence graph.

Table 2. Representative current methods from the reviewed literature and their relevance to TSA-LLM.

Area	Examples	Contribution to LLM time-step alignment
Text-aligned speech tokens	PAST, prosody-token benchmarking, TASTE, TASTE-Streaming, Latent Speech-Text Transformer, TADA [8,21,30,35,45,46]	Compress acoustic streams into LLM-compatible units while trying to preserve linguistic or temporal structure
LLM forced alignment and timestamps	LLM-ForcedAligner, In-Sync, Whisper internal aligner [13,33,50]	Use LLM-compatible or speech-aware models to produce word-level or transcript-level time spans
Timestamp-grounded reasoning	Listen First, Speaker-Reasoner, ALARM, TARS [18,25,29,47]	Evaluate whether generated answers and reasoning claims are actually grounded in localized speech evidence
Timestamped audio events and captions	TAC, CASTELLA [26,34]	Extend grounding from words to audio events, captions, and long-audio boundaries
Retrieval and contextual ASR	kNN-CTC, DYNAC, SEAL, CLAR [24,43,44,55]	Improve rare-entity handling but introduce provenance questions: heard, retrieved, or inferred?
Alignment preference and modality effects	Same Words Different Judgments, Speech-Worthy Alignment, integration-vs-combination analysis [5,41,53]	Show that speech LLM behavior is not reducible to text-only LLM behavior and that alignment preferences depend on modality

5 Current LLM-Oriented Methods in the Reviewed Literature

The most recent methods reviewed here move from ASR alignment toward speech LLM interfaces. Table 2 maps representative methods to their contribution for TSA-LLM.

5.1 Text-Aligned Speech Tokenization and Adapters

A speech LLM cannot usually pass frame-level features directly to a text LLM because the frame sequence is too long. It uses discrete tokens, continuous tokens, pooled encoder states, or adapter outputs. The alignment risk is that this compression may discard timing information. Self-supervised encoders such as wav2vec 2.0 and HuBERT supply strong frame-level representations [1,23], while speech–text pre-training methods such as SpeechUT and SpeechLM aim to bridge speech and text representations [51,52].

Recent token-interface methods make the alignment issue explicit. PAST emphasizes phonetic-acoustic tokenization [21]. Prosody-token benchmarking asks whether discrete speech tokens preserve prosodic information [35]. TASTE and TASTE-Streaming pursue text-aligned speech tokenization and streamable tokenization [45,46]. The Latent Speech-Text Transformer and TADA address latent or dual alignment between speech and text [8,30]. The TSA-LLM question for all of these methods is whether the resulting token sequence carries recoverable time provenance. A token may be semantically aligned with text while still being temporally opaque.

5.2 LLM-Based Forced Alignment and Word Timestamp Prediction

LLM-based forced alignment and timestamp prediction bring large-model semantic priors into alignment. LLM-ForcedAligner casts forced alignment in a form designed for multilingual and long-form settings [33]. In-Sync adapts speech-aware LLMs for ASR with word-level timestamp predictions [13]. These methods are a natural bridge between ASR aligners and speech LLMs: they provide explicit spans, but they also use richer language conditioning.

The key risk is timestamp hallucination. LLMs are good at generating plausible numeric strings and citation formats. A TSA-LLM system should therefore separate symbolic generation from boundary selection whenever possible. The LLM can request or emit a citation slot, while a calibrated boundary decoder, CTC posterior, transducer lattice, CIF localization, or span head fills the numeric interval. This separation turns timestamps from generated text into auditable model outputs.

5.3 Timestamp-Grounded Reasoning, Captioning, and Attribution

Reasoning tasks reveal failures that word timestamps alone cannot detect. Listen First frames speech reasoning around timestamp-grounded evidence [25]. Speaker-Reasoner adds speaker attribution and interaction patterns to timestamped ASR [29]. ALARM and TARS study audio-language alignment and modality reasoning for reasoning models and speech LLMs [18,47]. These methods shift the unit of evaluation from word boundary to evidence use.

Timestamped audio captioning and long-audio caption datasets extend the same idea to non-ASR outputs. TAC targets timestamped audio captioning, and CASTELLA supplies long audio with captions and temporal boundaries [26,34]. These tasks matter for speech LLMs because user-facing systems often need to cite events, acoustic contexts, or speaker actions, not only transcript words.

5.4 Retrieval, Contextual Biasing, and Rare Entities

Rare words, names, and technical terms stress both recognition and grounding. Contextual biasing and retrieval can improve recognition, but they make provenance ambiguous. kNN-CTC retrieves CTC pseudo-label information [55]; DYNAC performs dynamic vocabulary-based contextualization [43]; SEAL aligns speech embeddings for retrieval-augmented speech LLMs [44]; and CLAR uses CIF-localized alignment for retrieval-augmented contextual ASR [24]. A correct entity output is not fully grounded unless the system states whether the entity was heard in audio, retrieved from external context, or inferred from the prompt.

5.5 Multilingual, Phonetic, and Pronunciation-Oriented Alignment

Multilingual and phonetic methods broaden TSA-LLM beyond English word timestamps. LLM-based phoneme-to-grapheme work, universal phone recognition, CTC-based dialect identification, pronunciation scoring, and phonetic transcription robustness all address relations among phones, graphemes, languages, and acoustic evidence [4,6,11,14,49]. For multilingual speech LLMs, the alignment object must sometimes attach to phones or pronunciations rather than surface words. This is especially important for code switching, low-resource languages, non-standard pronunciations, and accent-sensitive applications.

6 The Alignment Contract

Across the surveyed methods, we identify a common design need: a speech LLM should return an alignment contract, not just text and optional timestamps. The contract is a public interface for each grounded output unit. It specifies role, provenance, audio support, uncertainty, evidence-token identifiers, and commitment status. It can be implemented with different architectures: CTC scaffolds, transducer lattices, CIF tokenization, learned adapters, LLM timestamp heads, retrieval modules, or post-hoc aligners.

6.1 Evidence Tokens with Time Provenance

Let $\alpha_{i,m}$ be the contribution of encoder state m to evidence token i, with $\sum_m \alpha_{i,m} = 1$. A general evidence token can be written as

$$v_i = W_v \sum_{m=1}^{M} \alpha_{i,m} h_m + e^{\text{time}}(\mu_i) + e^{\text{role}}(\rho_i), \tag{3}$$

where $\mu_i = \sum_m \alpha_{i,m} m$ is a soft time center and ρ_i is a role prior. The important interface requirement is not this exact formula, but the preservation of α or a compressed equivalent after tokenization. Without it, the LLM can use speech tokens semantically but cannot trace them back to audio time.

6.2 Span and Role Estimation

For a generated step k with hidden state g_k, a span estimator can relate the step to evidence tokens:

$$\beta_{k,i} = \mathrm{softmax}_i(g_k^\top W_b v_i), \qquad q_k(m) = \sum_{i=1}^{N} \beta_{k,i} \alpha_{i,m}. \tag{4}$$

The distribution $q_k(m)$ can be summarized as one interval, multiple intervals, or an uncertainty band. A role classifier predicts whether the step is lexical evidence, speaker evidence, prosodic evidence, acoustic-event evidence, retrieved context, instruction following, formatting, or ungrounded prior. Role prediction prevents artificial audio spans from being assigned to every token.

6.3 Constrained Timestamp Generation

Many systems ask an LLM to emit timestamp strings. A safer approach is constrained timestamp generation: the LLM generates a citation request, while a boundary model chooses numeric times from calibrated candidates. Candidate sources can include CTC posteriors, transducer alignments, CIF firing weights, forced-aligner outputs, Whisper-style internal alignments, and learned span heads. This design connects LLM-based forced aligners and speech-aware timestamp heads to timestamp-grounded reasoning, while reducing free-form timestamp hallucination.

6.4 Streaming Commitments

Streaming systems need commitment states. A provisional span is an early estimate that may be revised. A committed span is sufficiently supported to cite. A revised span records a correction. The delay metric should be attached to commitment, not only to token emission. For example, a model may emit an answer token at time t but not be able to cite the evidence until $t + \Delta$. Reporting Δ is crucial for live captioning, spoken dialogue, and safety-critical audio agents.

7 Training and Supervision Regimes

A TSA-LLM system can combine multiple objectives:

$$\begin{aligned}\mathcal{L} = \mathcal{L}_{\mathrm{LM}} + \lambda_{\mathrm{ctc}}\mathcal{L}_{\mathrm{ctc}} + \lambda_{\mathrm{span}}\mathcal{L}_{\mathrm{span}} + \lambda_{\mathrm{role}}\mathcal{L}_{\mathrm{role}} \\ + \lambda_{\mathrm{mono}}\mathcal{L}_{\mathrm{mono}} + \lambda_{\mathrm{loc}}\mathcal{L}_{\mathrm{loc}} + \lambda_{\mathrm{lat}}\mathcal{L}_{\mathrm{lat}} + \lambda_{\mathrm{cf}}\mathcal{L}_{\mathrm{cf}}.\end{aligned} \tag{5}$$

The language-modeling term supports semantic generation, but it does not guarantee correct timestamps. Posterior alignment terms supply monotonic frame-label supervision. Span and role losses teach explicit supports when labels exist. Monotonicity and locality regularizers discourage diffuse assignments. Latency terms penalize late commitment. Counterfactual terms test whether cited evidence is actually used.

7.1 Timestamped, Transcript-Only, and Text-Only Data

Timestamped transcripts can train word or subword spans directly. Transcript-only data can train CTC or transducer scaffolds and provide weak path marginals. Text-only alignment simulation and low-resource adaptation methods such as TASU and TASU2 illustrate how alignment supervision can be constructed when speech labels are limited [36,37]. For speech LLMs, these weak sources may be necessary because claim-level timestamped reasoning data remains scarce.

7.2 Reasoning and Citation Data

Reasoning data changes the target from word boundaries to evidence supports. An answer should be aligned to the audio regions that justify it, and a speaker claim should be aligned to both words and speaker identity. Negative controls are essential: a model should not receive credit for citing a span that contains no evidence. Timestamp-grounded reasoning, speaker-reasoning, and audio-language reasoning alignment datasets are therefore important complements to forced-alignment benchmarks [18,25,29,47].

7.3 Counterfactual Faithfulness

A citation is faithful only if the cited audio matters. Let z be a generated claim and S_z its cited support. A counterfactual objective can penalize the model when masking the cited span does not reduce the probability of the claim:

$$\mathcal{L}_{\text{cf}} = \max\left(0, \gamma - \log p_\theta(z \mid x) + \log p_\theta(z \mid \text{mask}(x, S_z))\right). \tag{6}$$

The same idea can be implemented by span dropout, time shifting, audio replacement, hard negative spans, or contrastive evidence. This is a test of public-output faithfulness, not a request to expose private reasoning traces.

8 Evaluation Protocol

The survey conclusion is that TSA-LLM cannot be evaluated by one metric. A system can have high ASR accuracy and poor grounding; another can have good word timestamps and poor answer citations. We recommend reporting the following metric families.

8.1 Boundary and Span Accuracy

For word, phone, event, and claim spans, report mean absolute boundary error, median error, and tolerance-window accuracy at 20, 50, 100, and 200 ms where appropriate. For long events or claim supports, report temporal intersection-over-union, coverage, and precision. Forced-alignment benchmarks should be linked to speech LLM outputs rather than treated as isolated preprocessing [9].

8.2 Grounded Faithfulness

For answers, summaries, captions, and citations, corrupt cited spans and measure whether answer probability, answer consistency, or entailment changes. Corrupt non-cited spans as controls. A faithful model should be more sensitive to corruption of cited evidence than to corruption of unrelated audio. This evaluation is especially important for timestamp-grounded reasoning, speaker attribution, and retrieval-augmented ASR.

8.3 Calibration

Report calibration for boundary confidence, span confidence, and provenance confidence. If a model produces a 90% interval for a boundary, the true boundary should fall inside it approximately 90% of the time. If it gives citation confidence, bins of citation confidence should correlate with empirical temporal correctness and evidence usefulness.

8.4 Drift and Long-Form Stability

Long-form audio should be evaluated by cumulative drift over recording time. Systems may be accurate on short utterances but lose synchronization after segmentation, silence, overlapping speech, speaker changes, or domain shifts. Report drift at fixed offsets, robustness to segmentation boundaries, and behavior under speaker overlap. Long-form timestamping systems make these issues visible in practice [2, 38].

8.5 Latency and Revision

Streaming evaluation should report token emission delay, evidence-commitment delay, revision rate, real-time factor, memory, and update frequency. The latency of a speech LLM should not be measured only by first-token time. A grounded system must also state when the cited support is reliable enough to expose.

8.6 Error Stratification

Errors should be stratified by language, accent, speaking rate, overlap, noise, domain, rare-entity frequency, prompt type, and output role. Current work on robust CTC, accented recognition, dialect identification, universal phone recognition, pronunciation scoring, and modality-specific preference alignment suggests that alignment quality can vary across populations and task formats [3–6, 14, 28, 53].

9 Discussion

Surveying TSA-LLM reveals a central tension. ASR-derived alignment is temporally disciplined but often semantically shallow: it locates words, not necessarily answer evidence. LLM-derived timestamp generation is semantically flexible but can hallucinate times. The strongest systems are likely to be hybrids. CTC, transducer, CIF, and forced-alignment scaffolds provide temporal constraints; speech-token interfaces make audio consumable by LLMs; and LLM reasoning modules map localized evidence into answers, summaries, captions, or speaker claims. The missing piece is an auditable contract among these modules.

The survey also motivates a terminology shift. Temporal alignment should not be treated only as preprocessing for speech LLMs. It is a model-interface problem. A speech LLM that cannot expose evidence supports is difficult to debug, calibrate, or trust, even if its text output is correct. Conversely, a model that exposes typed, calibrated supports can be evaluated and improved along dimensions that ordinary language metrics miss.

10 Conclusion

This paper surveyed LLM-based time-step alignment for speech–text grounding. We defined TSA-LLM as the problem of aligning LLM conditioning and generation steps with explicit audio-time supports; reviewed classical scaffolds, forced aligners, speech-token interfaces, LLM-based timestamping, reasoning, captioning, retrieval, and multilingual methods from the reviewed literature; and synthesized them into an alignment-contract view. The main recommendation is that future speech LLMs should not merely generate text with optional timestamps. They should expose typed, uncertainty-aware, provenance-labeled supports that show which audio time steps justify their public outputs.

Acknowledgments. This work was supported by JST BOOST, Japan Grant Number JPMJBY25F6.

Disclosure of Interests. The authors declare no competing interests relevant to the content of this article.

References

1. Baevski, A., Zhou, Y., Mohamed, A., Auli, M.: wav2vec 2.0: a framework for self-supervised learning of speech representations. In: Advances in Neural Information Processing Systems, vol. 33, pp. 12449–12460 (2020)
2. Bain, M., Huh, J., Han, T., Zisserman, A.: WhisperX: time-accurate speech transcription of long-form audio. In: Proceedings of Interspeech, pp. 4489–4493 (2023). https://doi.org/10.21437/Interspeech.2023-78
3. Bartelds, M., Nandi, A., Doumbouya, M.K.B., Jurafsky, D., Hashimoto, T., Livescu, K.: CTC-DRO: robust optimization for reducing language disparities in speech recognition (2025). ICLR 2026

4. Bharadwaj, S., et al.: An empirical recipe for universal phone recognition (2026)
5. Broukhim, A., Weibel, N., Jolly, E.: Same words, different judgments: modality effects on preference alignment (2026)
6. Cao, X., Fan, Z., Svendsen, T., Salvi, G.: Segmentation-free goodness of pronunciation (2025)
7. Chan, W., Jaitly, N., Le, Q.V., Vinyals, O.: Listen, attend and spell. In: Proceedings of ICASSP, pp. 4960–4964 (2016). https://doi.org/10.1109/ICASSP.2016.7472621
8. Dang, T., et al.: Tada: a generative framework for speech modeling via text-acoustic dual alignment (2026)
9. Deng, Y., Richardson, F., Steinberg, J.: Speech-to-text forced alignment benchmark. In: Proceedings of the 14th International Symposium on Image and Signal Processing and Analysis, pp. 145–150 (2025). https://doi.org/10.1109/ISPA66905.2025.11259471
10. Dong, L., Xu, S., Xu, B.: CIF: continuous integrate-and-fire for end-to-end speech recognition. In: Proceedings of ICASSP, pp. 6079–6083 (2020). https://doi.org/10.1109/ICASSP40776.2020.9053845
11. Dong, L., Li, Z., Yusuyin, S., Zhao, X., Ou, Z.: Advancing LLM-based phoneme-to-grapheme for multilingual speech recognition (2026)
12. Eom, S., Yoon, E., Yoon, H.S., Kim, C., Hasegawa-Johnson, M., Yoo, C.D.: AdaMER-CTC: connectionist temporal classification with adaptive maximum entropy regularization for automatic speech recognition (2024)
13. Fan, X., Sunder, V., Thomas, S., Hasegawa-Johnson, M., Kingsbury, B., Saon, G.: In-sync: adaptation of speech aware large language models for ASR with word level timestamp predictions (2026)
14. Farooq, M.U., Saz, O.: CTC-did: CTC-based Arabic dialect identification for streaming applications (2026)
15. Grafé, H., Van hamme, H.: Graph connectionist temporal classification for phoneme recognition (2025)
16. Graves, A.: Sequence transduction with recurrent neural networks. In: ICML Workshop on Representation Learning (2012)
17. Graves, A., Fernández, S., Gomez, F., Schmidhuber, J.: Connectionist temporal classification: labelling unsegmented sequence data with recurrent neural networks. In: Proceedings of the 23rd International Conference on Machine Learning, pp. 369–376 (2006). https://doi.org/10.1145/1143844.1143891
18. Grinberg, P., Shahmohammadi, H.: Alarm: audio-language alignment for reasoning models (2026)
19. Gulati, A., et al.: Conformer: convolution-augmented transformer for speech recognition. In: Proceedings of Interspeech, pp. 5036–5040 (2020). https://doi.org/10.21437/Interspeech.2020-3015
20. Haliassos, A., Mira, R., Petridis, S.: Pay attention to CTC: fast and robust pseudo-labelling for unified speech recognition (2026). Published as a conference paper at ICLR 2026
21. Har-Tuv, N., Tal, O., Adi, Y.: Past: phonetic-acoustic speech tokenizer. In: Proceedings of Interspeech (2025)
22. Hilmes, B., Rossenbach, N., Schlüter, R.: Analyzing the importance of blank for CTC-based knowledge distillation (2025)
23. Hsu, W.N., Bolte, B., Tsai, Y.H.H., Lakhotia, K., Salakhutdinov, R., Mohamed, A.: HuBERT: self-supervised speech representation learning by masked prediction of hidden units. In: IEEE/ACM Transactions on Audio, Speech, and Language Processing, vol. 29, pp. 3451–3460 (2021). https://doi.org/10.1109/TASLP.2021.3122291

24. Huang, S., Shen, H., Zou, W., Chen, Y.: CLAR: CIF-localized alignment for retrieval-augmented speech LLM-based contextual ASR (2026)
25. Jeong, J., Mousavi, P., Ravanelli, M., Subakan, C.: Listen first, then answer: timestamp-grounded speech reasoning (2026)
26. Kumar, S., et al.: TAC: timestamped audio captioning (2026)
27. Le, K., Ho, T.V., Tran, D., Chau, D.T.: SegAug: CTC-aligned segmented augmentation for robust RNN-transducer based speech recognition (2025)
28. Lee, W., Kim, H., Lee, G.G.: Mixture-of-experts with intermediate CTC supervision for accented speech recognition (2026)
29. Lin, Z., et al.: Speaker-reasoner: scaling interaction turns and reasoning patterns for timestamped speaker-attributed ASR (2026)
30. Lu, Y.J., et al.: Latent speech-text transformer (2025)
31. McAuliffe, M., Socolof, M., Mihuc, S., Wagner, M., Sonderegger, M.: Montreal forced aligner: trainable text-to-speech alignment using Kaldi. In: Proceedings of Interspeech, pp. 498–502 (2017). https://doi.org/10.21437/Interspeech.2017-1386
32. Mo, J.H., Kim, D.H., Lee, J.H., Chang, J.H.: Alignment regularization for neural transducer via cross-modal optimal transport. IEEE Signal Process. Lett. (2024). Metadata verified from seed list and public indexing
33. Mu, B., Shi, X., Wang, X., Liu, H., Xu, J., Xie, L.: LLM-ForcedAligner: a non-autoregressive and accurate LLM-based forced aligner for multilingual and long-form speech (2026)
34. Munakata, H., Imamura, T., Nishimura, T., Komatsu, T.: Castella: long audio dataset with captions and temporal boundaries (2025)
35. Onda, K., Fukayama, S., Saito, D., Minematsu, N.: Benchmarking prosody encoding in discrete speech tokens (2025)
36. Peng, J., et al.: TASU2: controllable CTC simulation for alignment and low-resource adaptation of speech LLMs (2026)
37. Peng, J., et al.: TASU: text-only alignment for speech understanding (2025)
38. Radford, A., Kim, J.W., Xu, T., Brockman, G., McLeavey, C., Sutskever, I.: Robust speech recognition via large-scale weak supervision (2023)
39. Raffel, C., Luong, M.T., Liu, P.J., Weiss, R.J., Eck, D.: Online and linear-time attention by enforcing monotonic alignments. In: Proceedings of ICML, pp. 2837–2846 (2017)
40. Rehman, A., Cai, J., Zhang, J.J., Yang, X.: BFA: real-time multilingual text-to-speech forced alignment (2025)
41. Schmitt, R., Zeyer, A., Zeineldeen, M., Schlüter, R., Ney, H.: LLMs and speech: integration vs. combination (2026)
42. She, Q., Peng, J., Fang, Y., Xi, Y., Yu, K.: TC-BiMamba: trans-chunk bidirectionally within BiMamba for unified streaming and non-streaming ASR (2026)
43. Sudo, Y., Fukumoto, Y., Shakeel, M., Peng, Y., Lin, C.J., Watanabe, S.: DYNAC: dynamic vocabulary-based non-autoregressive contextualization for speech recognition. In: Proceedings of Interspeech, pp. 2215–2219 (2025). https://doi.org/10.21437/Interspeech.2025-129
44. Sun, C., et al.: SEAL: speech embedding alignment learning for speech large language model with retrieval-augmented generation (2025)
45. Tseng, L.H., Chen, Y.C., Lee, K.Y., Shiu, D.S., Lee, H.Y.: TASTE: text-aligned speech tokenization and embedding for spoken language modeling (2025)
46. Tseng, L.H., Lee, H.Y.: Taste-streaming: towards streamable text-aligned speech tokenization and embedding for spoken language modeling (2026)
47. Wang, C., et al.: Closing the modality reasoning gap for speech large language models (2026)

48. Yao, Z., et al.: Delay-penalized CTC implemented based on finite state transducer. In: Proceedings of Interspeech (2023)
49. Ye, Z., et al: LCS-CTC: leveraging soft alignments to enhance phonetic transcription robustness (2025)
50. Yeh, S.L., Meng, Y., Tang, H.: Whisper has an internal word aligner (2025)
51. Zhang, Z., et al.: SpeechLM: enhanced speech pre-training with unpaired textual data. IEEE/ACM Trans. Audio Speech Lang. Process. (2024). https://doi.org/10.1109/TASLP.2024.3379877
52. Zhang, Z., et al.: SpeechUT: bridging speech and text with hidden-unit for encoder-decoder based speech-text pre-training. In: Proceedings of EMNLP, pp. 1663–1676. Association for Computational Linguistics (2022)
53. Zhao, M., et al.: Speech-worthy alignment for Japanese SpeechLLMs via direct preference optimization (2026)
54. Zheng, Y., Mei, Y., Xu, D., Chen, J., Long, Y.: A language-agnostic hierarchical LoRA-MoE architecture for CTC-based multilingual ASR (2026)
55. Zhou, J., Zhao, S., Liu, Y., Zeng, W., Chen, Y., Qin, Y.: KNN-CTC: enhancing ASR via retrieval of CTC pseudo labels. In: Proceedings of ICASSP (2024). https://doi.org/10.1109/ICASSP48485.2024.10447075

Affective Interaction and Social Computing

Richness-Aware Modality-Role Assignment for Multimodal Emotional Dialogue Generation

Mingting Yu, Changhong Jiang(✉), Zhongtian Hu, Wei Zhang, and Lulu Wang

Northwestern Polytechnical University, Xi'an 710129, China
{mingtingyu,ahxchzt,zhang_wei,wangllu}@mail.nwpu.edu.cn,
chjiang@nwpu.edu.cn

Abstract. Multimodal emotional dialogue generation (EDG) has benefited from increasingly powerful fusion architectures, yet many systems still adopt a text-centric and statically fused multimodal pipeline and fail to fully exploit audio–visual affective cues whose salience varies across dialogue contexts. We present **RaMRA**, a *richness-aware modality-role assignment* framework that estimates the semantic–sentiment informativeness of text, audio, and vision and dynamically assigns instance-level dominant/auxiliary roles. The dominant stream serves as an interaction hub for role-conditioned cross-modal attention, while response generation remains anchored in the textual context. Experiments on IEMOCAP and MELD demonstrate that RaMRA consistently improves fluency, semantic alignment, and emotion consistency over strong baselines, underscoring the importance of adaptive multimodal affect perception for building emotionally intelligent dialogue agents.

Keywords: Multimodal dialogue generation · Modality-role assignment · Role-aware fusion · Affective computing · Emotional intelligence

1 Introduction

Dialogue generation has become a core interface for intelligent customer service, virtual assistants, and everyday human–computer interaction. While neural generators increasingly master coherence and contextual relevance, they still tend to under-model affect, resulting in responses that sound fluent yet emotionally off-target [8]. Emotional dialogue generation (EDG) therefore seeks to produce replies that are not only contextually appropriate but also aligned with the speaker's affective state—a capability at the heart of *emotional intelligence* in human–AI interaction and a prerequisite for truly empathetic dialogue agents.

Multimodal EDG is often motivated by a simple observation: the same words can convey different emotions under different prosody or facial display. Recent systems accordingly incorporate audio and vision [3,6], yet many pipelines

H. Chen et al. (Eds.): IJCAI 2026, CCIS 3120, pp. 187–200, 2027.
https://doi.org/10.1007/978-981-92-4529-1_13

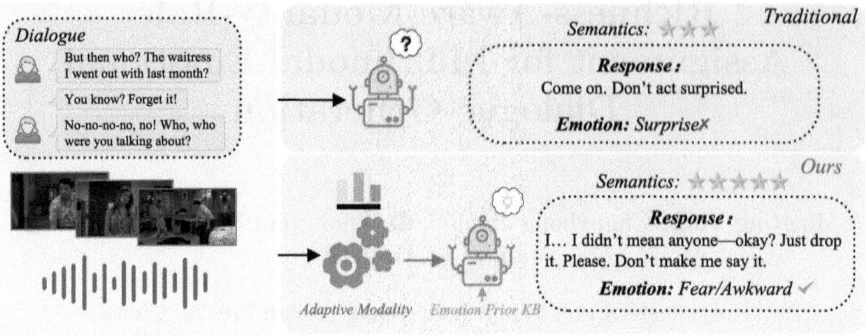

Fig. 1. Motivation and overview of **RaMRA**. Conventional multimodal EDG is often text-centric, which can overlook context-dependent affective cues. The Modality-Role Discriminator and Role-Aware Fusion Module jointly assign primary/auxiliary roles and route fusion through the primary modality, while an *Emotion Prior KB* injects affective priors to improve emotion-aligned generation.

remain *text-centric* in a structural sense—text acts as the default backbone, and non-text streams are appended as auxiliary refinements or global conditions. This design implicitly assumes that extra modalities are uniformly helpful, which is rarely true in practice: acoustic cues may dominate when wording is neutral but tone is tense, while facial expressions can be decisive when speech is short or ambiguous. Even systems built around a *fixed* non-text primary modality inherit the same limitation—the weighting prior is predetermined rather than inferred per instance, making them brittle under modality-specific noise, missing cues, or cross-modal incongruence [6,12,22]. This motivates a key question: "can multimodal EDG explicitly decide which modality should lead the interaction for each dialogue context, while keeping generation anchored to textual discourse?"

As illustrated in Fig. 1, we argue that multimodal EDG should separate two functional roles: (i) a *grounding channel* that preserves dialogue semantics and controllability (naturally, text), and (ii) an *interaction hub* that adaptively routes cross-modal affective evidence to maximize emotional reliability. To instantiate this principle, we propose **RaMRA**[1], an informativeness-driven modality-role assignment and role-conditioned interaction framework within a decoder-only generation pipeline. For each dialogue context, RaMRA estimates the semantic–sentiment informativeness of text, audio, and vision, assigns a dominant/auxiliary role configuration, and performs cross-modal exchange *asymmetrically*: the dominant modality serves as the sole query source for cross-modal attention, while auxiliary streams contribute evidence only through value vectors modulated by their informativeness weights. This role-based asymmetry distinguishes RaMRA from prior dynamic fusion approaches [14,15,21,24], which reweight modality contributions through *scalar* gates yet still maintain symmetric participation of every modality in attention. In RaMRA, high-quality signals

[1] RaMRA stands for **R**ichness-**a**ware **M**odality-**R**ole **A**ssignment.

are amplified through hub-based routing rather than averaged with noise from less reliable streams. Decoding remains anchored in textual discourse, and an auxiliary emotion prediction objective on decoder hidden states further tightens the coupling between affective understanding and response generation [8].

Our contributions are threefold:

1. **Role-based reformulation.** We formulate multimodal EDG as *instance-level modality-role assignment* and propose a richness-aware criterion that assigns roles based on semantic–sentiment informativeness.
2. **Role-based asymmetric fusion.** We introduce role-conditioned asymmetric interaction that routes cross-modal evidence via gated attention, reducing redundancy and interference from less informative streams.
3. **Framework & validation.** We present RaMRA as a plug-in framework on top of a decoder-only generator with frozen modality encoders, and validate its effectiveness on IEMOCAP and MELD through analyses and ablations.

2 Related Work

Emotional Dialogue Generation. Emotional dialogue generation (EDG) aims to produce responses that are both contextually coherent and emotionally aligned with the speaker's affective state. Text-only approaches have progressively enriched affective grounding through commonsense knowledge infusion [9], emotion-sensitive recognition [13], coarse-to-fine cognition–affection alignment [27], and iterative associative memory [23]. A closely related upstream task, multimodal emotion recognition in conversations (ERC), provides the perceptual foundation that EDG systems rely on; recent ERC advances include attention-based correlation-aware fusion [10], adaptive graph learning [11], and graph-spectral analysis of cross-modal emotion dynamics [1], as surveyed in [16].

Building on these recognition capabilities, multimodal EDG systems have begun to incorporate audio and visual signals alongside text. Representative efforts include the open-source avatar chatbot EmpathyEar [3], the emotion-guided LLM pipeline SpeakFromHeart [6], multi-modality fusion in ERGM [18], LLM-driven empathetic speech generation in EmpathyOmni [12], the text-speech-vision avatar benchmark AvaMERG [25], and personality-aware multimodal empathetic generation [17]. Despite their progress, these systems share a common architectural assumption: either text serves as the fixed backbone with non-text streams appended as auxiliary signals, or a single non-text modality is pre-designated as the primary channel. Neither design adapts to the *instance-level* variation in modality salience that naturally arises in emotional conversations.

Dynamic and Non-uniform Multimodal Fusion. Standard multimodal fusion strategies—such as concatenation, element-wise gating, and symmetric cross-attention—treat all modalities as equally informative, which can amplify noise from unreliable streams and dilute discriminative affective cues [4]. To address this, Dynamic Multimodal Fusion [21] estimates per-modality reliability at test

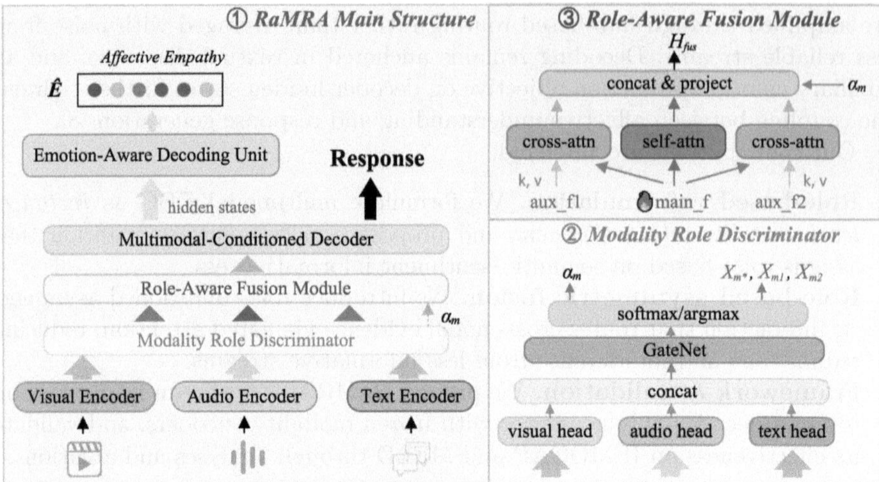

Fig. 2. Architecture of **RaMRA**, organized in three panels read in the order (1)→(2)→(3) to mirror the data flow. **(1) RaMRA Main Structure** (left): the end-to-end pipeline where frozen encoders feed the Modality-Role Discriminator (MRD), whose role weights α_m drive the Role-Aware Fusion Module (RAF), and an Emotion-Aware Decoding Unit supplies auxiliary affective supervision on decoder hidden states. Panels (2) and (3) zoom into the two highlighted submodules of (1): **(2) MRD** (bottom-right) produces α_m via gated tri-tower scoring, and **(3) RAF** (top-right) performs asymmetric routing where the dominant stream serves as the sole query source for cross-attention over auxiliaries, yielding $\mathbf{H}_{\text{fus}}$ for the decoder in (1).

time and adjusts contribution weights accordingly; HCMS [15] introduces hierarchical and conditional modality selection for efficient video recognition; and pre-gating with contextual attention gating [24] provides learnable mechanisms to suppress noisy channels before fusion. More recent work explores adaptive modality weighting through multimodal transformers [14] and multi-loss optimization across heterogeneous feature encoders [19], both targeting sentiment analysis. While these methods adaptively reweight modality contributions, they typically apply scalar adjustments that keep every modality symmetrically engaged in fusion, which is precisely the limitation RaMRA addresses through explicit dominant/auxiliary role assignment and asymmetric, role-conditioned routing (as outlined in Sect. 1).

3 Methodology

Task Definition. We study multimodal emotional dialogue generation with aligned text, audio, and vision. Given a dialogue context, we generate a contextually appropriate and emotion-aligned response; emotion labels are used only for training supervision and are not available at inference. Figure 2 overviews

RaMRA, which assigns modality roles to route cross-modal evidence while keeping decoding grounded in text.

3.1 Modality-Role Discriminator

Let $\mathcal{M} = \{T, A, V\}$ denote *text*, *audio*, and *vision*. Given a dialogue context, modality-specific encoders map raw inputs to modality-level embeddings:

$$\mathbf{x}_m = E_m(\text{input}_m), \quad m \in \mathcal{M}, \tag{1}$$

where E_T, E_A, E_V are fixed feature extractors. RaMRA operates exclusively on $\{\mathbf{x}_m\}$, avoiding direct dependence on raw waveforms or frames.

Tri-Tower Richness Scoring with Gated Calibration. We adopt a tri-tower parameterization to compute modality-specific latent summaries:

$$\mathbf{h}_m = \phi\big(\text{LN}(\mathbf{W}_m^{(h)}\mathbf{x}_m + \mathbf{b}_m^{(h)})\big), \quad m \in \mathcal{M}, \tag{2}$$

where $\phi(\cdot)$ is a nonlinearity and LN$(\cdot)$ stabilizes representation scales. To suppress unreliable dimensions, we introduce a learnable multiplicative gate:

$$\mathbf{g}_m = \sigma\big(\mathbf{W}_m^{(g)}\mathbf{x}_m + \mathbf{b}_m^{(g)}\big), \tag{3}$$

$$\tilde{\mathbf{h}}_m = \mathbf{g}_m \odot \mathbf{h}_m, \tag{4}$$

with sigmoid $\sigma(\cdot)$ and Hadamard product $\odot$. We then derive a scalar informativeness score via an affine readout:

$$s_m = \mathbf{u}_m^\top \tilde{\mathbf{h}}_m + c_m, \tag{5}$$

and convert scores into a continuous role distribution using a temperature-scaled normalization:

$$\alpha_m = \frac{\exp(s_m/\tau)}{\sum_{m' \in \mathcal{M}} \exp(s_{m'}/\tau)}, \quad m \in \mathcal{M}, \tag{6}$$

where τ controls the sharpness of the role distribution.

We define *informativeness* (or *richness*) operationally as the expected contribution of a modality to the joint objectives of semantic coherence and sentiment prediction in the current dialogue context. Concretely, for each modality m, the gated latent summary $\tilde{\mathbf{h}}_m$ (Eq. 4) isolates dimensions that carry discriminative affective and semantic value by multiplicatively suppressing uninformative channels; the scalar score s_m (Eq. 5) then aggregates this filtered signal into a single informativeness estimate. The entire pathway—from modality encoding through gating to scoring—is learned end-to-end under the combined language modeling and emotion classification losses (Eq. 15), so that a high s_m reflects evidence that is simultaneously useful for generating fluent, coherent text and for predicting the speaker's emotion. This joint optimization ensures that informativeness is not a hand-crafted heuristic but an empirically grounded quantity aligned with the downstream task.

We interpret $\boldsymbol{\alpha} = (\alpha_T, \alpha_A, \alpha_V)$ as instance-level role signals: the modality with the largest weight is the *dominant* stream, and the remaining modalities serve as auxiliary evidence providers.

3.2 Role-Aware Fusion Module

Conditioned on $\boldsymbol{\alpha}$, the fusion module integrates multimodal cues by explicitly routing cross-modal interactions through the dominant stream, rather than applying uniform aggregation.

Tokenized Descriptors. Let $\mathbf{X}_m \in \mathbb{R}^{L_m \times H}$ be the sequence-like representation of modality m after projecting all modalities to a shared hidden size H (e.g., via linear adapters). Here L_T is the number of text tokens, while L_A and L_V are pseudo-token lengths for acoustic/visual descriptors.

Role-Conditioned Routing. We define the dominant modality index

$$m^\star = \arg\max_{m \in \mathcal{M}} \alpha_m, \tag{7}$$

and the auxiliary set $\mathcal{M}_{\neg m^\star} = \mathcal{M} \setminus \{m^\star\}$. The full distribution $\boldsymbol{\alpha}$ is retained for continuous weighting, while $m^\star$ specifies the interaction hub. Although arg max introduces a discrete selection, cross-modal evidence from auxiliary modalities is integrated through *continuous* weights α_m (Eq. 6), so near-tie cases (e.g., $\alpha_A = 0.49$ vs. $\alpha_V = 0.51$) yield only marginal changes in the fused representation when the hub identity swaps; the temperature τ additionally keeps the role distribution soft near decision boundaries.

Intra-stream Refinement. We first apply self-attention to the dominant stream:

$$\mathbf{S}_{m^\star} = \mathrm{Attn}\Big(\mathbf{X}_{m^\star}\mathbf{W}^{Q}_{m^\star}, \mathbf{X}_{m^\star}\mathbf{W}^{K}_{m^\star}, \mathbf{X}_{m^\star}\mathbf{W}^{V}_{m^\star}\Big), \tag{8}$$

Cross-Modal Integration. For each auxiliary modality $m \in \mathcal{M}_{\neg m^\star}$, we compute dominant-query cross-attention:

$$\mathbf{C}_{m \to m^\star} = \mathrm{Attn}\Big(\mathbf{X}_{m^\star}\mathbf{W}^{Q}_{m^\star},\ \mathbf{X}_{m}\mathbf{W}^{K}_{m},\ \mathbf{X}_{m}\mathbf{W}^{V}_{m}\Big), \tag{9}$$

and modulate it by the corresponding role weight:

$$\widehat{\mathbf{C}}_{m \to m^\star} = \alpha_m \cdot \mathbf{C}_{m \to m^\star}. \tag{10}$$

Here $\mathrm{Attn}(\mathbf{Q}, \mathbf{K}, \mathbf{V}) = \mathrm{softmax}\big(\mathbf{Q}\mathbf{K}^\top / \sqrt{d}\big)\mathbf{V}$ (multi-head in practice).

Fusion Output and Decoding Interface. We concatenate the dominant stream, its refined state, and all weighted cross-modal evidence:

$$\mathbf{Z} = \Big[\mathbf{X}_{m^\star} \,\|\, \mathbf{S}_{m^\star} \,\|\, \{\widehat{\mathbf{C}}_{m \to m^\star}\}_{m \in \mathcal{M}_{\neg m^\star}}\Big], \tag{11}$$

and map it back to the decoder hidden size:

$$\mathbf{H}_{\mathrm{fus}} = \mathrm{LN}(\mathbf{Z}\mathbf{W}_{\mathrm{fus}} + \mathbf{b}_{\mathrm{fus}}). \tag{12}$$

To preserve textual grounding, we feed the decoder with text input embeddings augmented by $\mathbf{H}_{\mathrm{fus}}$ (via concatenation followed by a lightweight projection), rather than allowing any non-text stream to supplant the textual backbone.

3.3 Emotion-Aware Decoding Unit

To couple response generation with affective modeling, we attach an Emotion-Aware Decoding Unit to decoder hidden states and optimize it jointly with language modeling. Let $\mathbf{h}_T \in \mathbb{R}^H$ denote the decoder hidden state at the last response position under teacher forcing.[2] We compute emotion logits:

$$\mathbf{z}^{(e)} = \mathbf{W}_e \,\mathrm{LN}(\mathbf{h}_T) + \mathbf{b}_e, \quad \mathbf{p}^{(e)} = \mathrm{softmax}(\mathbf{z}^{(e)}), \tag{13}$$

and define the auxiliary objective with cross-entropy:

$$\mathcal{L}_{\text{emo}} = -\log \mathbf{p}^{(e)}\left[y^{(e)}\right]. \tag{14}$$

The final training objective is

$$\mathcal{L} = \mathcal{L}_{\text{lm}} + \mathcal{L}_{\text{emo}}, \tag{15}$$

which promotes emotion-discriminative decoder representations and supports emotion-aligned response generation.

4 Experiments

4.1 Datasets

We evaluate on **MELD** [7] and **IEMOCAP** [2], two widely used multimodal emotion benchmarks with aligned text, audio, and visual streams and utterance-level emotion labels. MELD contains 13,708 utterances from 1,433 dialogues annotated with 7 emotion categories, and IEMOCAP comprises about 7,433 utterances from 151 dyadic sessions with 6 commonly-used emotion categories. For MELD, we follow the official train/dev/test split. For IEMOCAP, we adopt the standard session-based protocol (Sessions 1–4 for training, Session 5 for testing) and hold out a validation subset from the training sessions for model selection. While leave-one-session-out cross-validation is the predominant protocol in the ERC literature [2], we opt for a fixed split primarily for two reasons: (1) it ensures a consistent comparison surface with MELD and with prior multimodal EDG work that also uses fixed train/test partitions [18]; and (2) IEMOCAP contains only five sessions, making each fold in a leave-one-out scheme a single session, which risks high evaluation variance on a test set of approximately 1,500 utterances and complicates checkpoint selection under heterogeneous session characteristics.

4.2 Implementation Details

All experiments run on an NVIDIA RTX 5090 GPU. Following ERGM [18], audio tracks are extracted with FFmpeg (16 kHz), word-aligned via the Montreal

[2] In our implementation, $\mathbf{h}_T$ corresponds to `last_token_hidden_state`.

Forced Aligner, and each utterance is segmented into a synchronized audio–visual clip of 32 frames. We use CLIP ViT-L/14 and HuBERT-Large as frozen visual and acoustic encoders, respectively. All modalities are projected to a shared hidden size of $H = 768$ via linear adapters, and multi-head attention uses 12 heads across all fusion modules.

For RaMRA and the ERGM baseline, both based on a decoder-only GPT-style backbone (12 layers, 12 heads, $H = 768$), we fine-tune directly on MELD and IEMOCAP using the Adam optimizer with a learning rate of 1×10^{-5}, a batch size of 16, and 10 training epochs; generation uses nucleus sampling with $p = 0.8$. For LLM-based settings, we use the Qwen2.5-Omni Thinker module with a learning rate of 1×10^{-4}, feed the concatenated dialogue history as input, keep modality encoders frozen, and apply LoRA ($r = 16$, $\alpha = 32$). Identical decoding hyperparameters are used across all LLM-based models. Other baselines (CEM, SEEK, CASE, IAMM) follow the hyperparameters in their original papers, and all baselines are reproduced using their official codebases. The ERGM baseline is reproduced without the guidance-text branch (which is not released) and without EmpatheticDialogues pre-training, matching the input pipeline available to RaMRA.

4.3 Evaluation Metrics

We report the following automatic metrics: perplexity (PPL) for fluency, Dist-1/Dist-2 [5] for response-level lexical diversity, BERTScore [26] (F1) for semantic similarity with the ground-truth response, and emotion accuracy (**Emo Acc.**), computed by a fixed external emotion classifier—an MLP head trained on each dataset's utterance-level emotion labels following the standard ERC protocol—applied to the generated responses. All models are evaluated by the same external classifier. We acknowledge that unweighted average recall (UAR) is often preferred over accuracy for imbalanced emotion datasets. However, we report accuracy for two reasons: (1) all baselines we compare against [9,13,18,23,27] use accuracy as their primary emotion metric, making it the de facto standard in multimodal EDG; and (2) the class distributions of both MELD (7 categories, 4,710 train utterances) and IEMOCAP (6 categories, 5,810 train utterances) are not severely skewed—the most and least frequent classes differ by a factor of approximately 3–4×, well within ranges where accuracy remains a reliable indicator. We additionally note that because all models are evaluated by the *same* external classifier under *identical* conditions, relative ranking under accuracy should closely mirror that under UAR.

Human Evaluation. We conduct a pairwise human study on 100 randomly sampled multimodal test instances from MELD and IEMOCAP (pooled for annotation). For each instance, three annotators with NLP research backgrounds compare two anonymized responses (order randomized) from RaMRA and the strongest baseline given the same multimodal inputs, and choose the better one under **Emp** (empathy/emotion alignment), **Coh** (contextual coherence), and **Flu** (fluency/naturalness). We report RaMRA's win rate per criterion. For

LLM-based human evaluation, we use the same format-only prompt at inference for both ERGM(Qwen) and RaMRA(Qwen).

5 Results

5.1 Automatic Evaluation

Table 1. Main results on MELD and IEMOCAP. Lower is better for PPL; higher is better for Dist-1/2, Emo Acc., and BERTScore. Best results are in **bold**.

	MELD					IEMOCAP				
Model	PPL↓	Dist-1↑	Dist-2↑	Emo Acc.↑	BERT↑	PPL↓	Dist-1↑	Dist-2↑	Emo Acc.↑	BERT↑
Standard models										
CEM [9]	53.68	0.91	4.28	58.40	76.59	41.03	0.97	3.62	61.87	76.21
SEEK [13]	51.44	1.01	5.38	58.95	81.01	36.52	1.12	3.81	66.40	82.01
CASE [27]	34.02	1.10	5.81	60.48	80.73	29.51	1.16	4.11	65.94	82.44
IAMM [23]	33.94	1.08	6.55	58.72	82.99	30.68	1.42	5.09	66.78	83.10
ERGM [18]	27.63	1.98	9.87	66.45	84.98	22.34	**3.16**	10.83	72.05	83.18
RaMRA (Ours)	**24.45**	**1.99**	**10.51**	**70.44**	**85.05**	**16.60**	**3.16**	**11.11**	**77.93**	**83.32**
LLM-based models										
Llama 3.2–11B	18.36	3.06	20.03	60.18	81.01	12.26	3.95	21.89	60.52	78.12
ERGM (Qwen2.5-Omni) [20]	17.61	3.44	21.00	67.45	83.31	12.07	**3.99**	22.09	63.37	80.19
RaMRA (Qwen2.5-Omni)	**16.96**	**3.48**	**22.57**	**68.89**	**86.93**	**10.74**	3.96	**22.31**	**65.25**	**86.24**

Standard Models. As shown in Table 1, RaMRA consistently outperforms the strongest standard baseline ERGM on both MELD and IEMOCAP, with the most pronounced gains on fluency and affect alignment. On MELD, RaMRA reduces PPL from 27.63 to **24.45** and improves Emo Acc. from 66.45 to **70.44**, alongside a modest but consistent increase in BERTScore (84.98→85.05). Diversity is preserved at the compositional level (Dist-2: 9.87→**10.51**), while Dist-1 remains essentially unchanged (1.98→1.99), suggesting that role-conditioned routing amplifies informative cross-modal cues without introducing superficial lexical novelty.

On IEMOCAP, the gains are larger and better reflect RaMRA's motivation under heterogeneous modality salience: PPL drops substantially (22.34→**16.60**) and Emo Acc. increases by nearly six points (72.05→**77.93**). RaMRA also improves Dist-2 and BERTScore (10.83→**11.11**, 83.18→**83.32**), indicating better lexical composition and semantic fidelity. Dist-1 ties with ERGM (3.16 vs. **3.16**), which is consistent with our goal of controlled diversity rather than aggressive lexical perturbation.

LLM-Based Models. RaMRA remains beneficial for omni-modal LLMs. Compared with ERGM (Qwen2.5-Omni [20]), RaMRA (Qwen2.5-Omni) reduces PPL on both MELD and IEMOCAP (17.61→**16.96**; 12.07→**10.74**) and improves Emo Acc. (67.45→**68.89**; 63.37→**65.25**). Notably, RaMRA yields a substantial BERTScore gain, especially on IEMOCAP (80.19→**86.24**), suggesting that modality-role routing helps the LLM internalize multimodal evidence more effectively than uniform fusion; Dist-2 also increases on both datasets (21.00→**22.57**; 22.09→**22.31**). The tiny Dist-1 drop on IEMOCAP (3.99→3.96; absolute decrease of 0.03) is negligible given RaMRA's consistent gains in PPL, Emo Acc., and BERTScore. We select Qwen2.5-Omni and Llama 3.2–11B as representative open-source native multimodal LLMs because they support audio, vision, and text as native inputs, enabling fair, reproducible comparisons under identical fine-tuning conditions. While Llama 3.2–11B exhibits low PPL and high diversity, it lags behind Qwen-based omni-modal models in Emo Acc. and BERTScore, reinforcing that explicit audiovisual grounding is essential for affect-aware generation.

5.2 Human Evaluation

Table 2. Human preference win rate (%) of RaMRA over ERGM under three criteria: Emp (empathy/emotion alignment), Coh (contextual coherence), and Flu (fluency/naturalness).

Comparison	Emp.↑	Coh↑	Flu↑
RaMRA vs. ERGM	62.6	56.4	52.7
RaMRA (Qwen) vs. ERGM (Qwen)	54.1	57.1	60.8

Table 2 reports pairwise human preferences. Overall, RaMRA achieves higher win rates than ERGM across all three criteria (Emp/Coh/Flu). In the Standard setting, RaMRA's advantage is most pronounced on empathy alignment (62.6%), aligning with our motivation of amplifying the most informative affective modality. In the LLM-based setting (Qwen2.5-Omni), the win-rate profile shifts: the empathy gap narrows to 54.1% while fluency becomes the main beneficiary (60.8%). We attribute this to the strong affective prior already encoded in Qwen2.5-Omni's large-scale multi-modal pre-training—when such an omni-modal LLM serves as the backbone, much of the speaker's emotion is already perceivable on its own, leaving RaMRA's role-conditioned routing to contribute mainly at the surface level of fluency and coherence.

5.3 Ablation Study

Table 3 shows that each component of RaMRA is necessary for its overall gains. Removing the Modality-Role Discriminator (uniform roles) consistently

Table 3. Ablation results. MRD: Modality-Role Discriminator; RAF: Role-Aware Fusion Module; EDU: Emotion-Aware Decoding Unit. "w/o" removes the component; w/o MRD uses uniform role weights; w/o RAF replaces it with naive symmetric fusion; w/o EDU disables the auxiliary emotion supervision while keeping the rest of the pipeline intact. Emo Acc. is computed using a fixed external classifier on generated responses (independent of EDU, so Emo Acc. remains computable in all settings).

	MELD			IEMOCAP		
Setting	PPL↓	EmoAcc.↑	BERT↑	PPL↓	EmoAcc.↑	BERT↑
Full (RaMRA)	24.45	70.44	85.05	16.60	77.93	83.32
w/o MRD	26.81	68.72	83.79	17.98	75.16	82.88
w/o RAF	27.09	67.55	82.94	17.49	74.59	81.30
w/o EDU	25.63	62.80	84.55	17.07	72.89	82.24

degrades fluency and semantic fidelity (higher PPL and lower BERTScore), indicating the importance of instance-level role inference. Replacing role-aware routing with naive fusion yields the largest semantic drop, especially on IEMOCAP (BERTScore: 83.32→81.30) with clear Emo Acc. degradation, confirming that dominant-modality routing is more effective than uniform aggregation. Finally, removing the emotion-aware unit causes the largest loss in emotion consistency (Emo Acc. −7.64 on MELD; −5.04 on IEMOCAP), suggesting auxiliary affective supervision is critical for emotion-aligned generation.

6 Conclusion

We propose **RaMRA**, a framework that reframes multimodal EDG as instance-level modality-role assignment. By identifying a context-specific interaction hub and performing role-conditioned, asymmetric routing under textual grounding, RaMRA reduces redundant fusion and emphasizes the most reliable affective evidence. This principled role-based formulation, together with auxiliary emotion supervision, consistently improves emotion-aligned generation on MELD and IEMOCAP, demonstrating that adaptive multimodal affect perception is key to building emotionally intelligent dialogue systems.

Limitations and Future Work. While RaMRA demonstrates consistent gains, several limitations warrant discussion. **(1) Dataset scope.** Our evaluation is restricted to two dyadic English conversation datasets (MELD and IEMOCAP); validation on larger-scale, multi-party, or multilingual benchmarks would strengthen the generality claim. **(2) Encoder freezing.** We use frozen CLIP and HuBERT encoders, following the setup in ERGM [18]. Fine-tuning the acoustic encoder could potentially recover more affective signal from prosody, at the cost of additional computational overhead. **(3) Human evaluation scale.** Our human study involves 100 instances and three annotators, which is adequate for pairwise preference estimation but limits the statistical power for fine-grained per-category analysis; larger-scale annotations with more annotators would yield more robust subjective metrics. **(4) Modality role analysis.** The current work reports aggregated performance metrics but does not provide a breakdown of how often each modality is selected as dominant, or how role assignments correlate with ground-truth emotion categories. **(5) Robustness to degraded inputs.** We do not evaluate performance under missing or noisy modalities, a practical concern in real-world deployment. Future work will extend RaMRA along these axes, including per-emotion-category analysis, modality role visualizations, and evaluation under modality degradation scenarios.

Disclosure of Interests. The authors have no competing interests to declare that are relevant to the content of this article.

References

1. Ai, W., Zhang, F., Shou, Y., Meng, T., Chen, H., Li, K.: Revisiting multimodal emotion recognition in conversation from the perspective of graph spectrum. In: Proceedings of the AAAI Conference on Artificial Intelligence, vol. 39, pp. 11418–11426 (2025). https://doi.org/10.1609/aaai.v39i11.33242
2. Busso, C., et al.: IEMOCAP: Interactive emotional dyadic motion capture database. In: Proceedings of the Sixth International Conference on Language Resources and Evaluation (LREC). European Language Resources Association (ELRA), Marrakech, Morocco (2008). https://doi.org/10.1007/s10579-008-9076-6
3. Fei, H., Zhang, H., Wang, B., Liao, L., Liu, Q., Cambria, E.: EmpathyEar: An open-source avatar multimodal empathetic chatbot. In: Proceedings of the 62nd Annual Meeting of the Association for Computational Linguistics (Volume 3: System Demonstrations), pp. 61–71. Association for Computational Linguistics, Bangkok, Thailand (2024). https://doi.org/10.18653/v1/2024.acl-demos.7
4. Feng, X., Lin, Y., He, L., Li, Y., Chang, L., Zhou, Y.: Knowledge-guided dynamic modality attention fusion framework for multimodal sentiment analysis. In: Findings of the Association for Computational Linguistics: EMNLP 2024, pp. 14755–14766. Association for Computational Linguistics, Miami, Florida, USA (2024). https://doi.org/10.18653/v1/2024.findings-emnlp.865
5. Li, J., Galley, M., Brockett, C., Gao, J., Dolan, B.: A diversity-promoting objective function for neural conversation models. In: Proceedings of the 2016 Conference of the North American Chapter of the Association for Computational Linguistics: Human Language Technologies (NAACL-HLT), pp. 110–119. Association for Computational Linguistics, San Diego, California (2016). https://doi.org/10.18653/v1/N16-1014
6. Liu, C., et al.: Speak from heart: An emotion-guided LLM-based multimodal method for emotional dialogue generation. In: Proceedings of the 2024 International Conference on Multimedia Retrieval (ICMR), pp. 533–542. ACM, Phuket, Thailand (2024). https://doi.org/10.1145/3652583.3658104
7. Poria, S., Hazarika, D., Majumder, N., Naik, G., Cambria, E., Mihalcea, R.: MELD: A multimodal multi-party dataset for emotion recognition in conversations. In: Proceedings of the 57th Annual Meeting of the Association for Computational Linguistics, pp. 527–536. Association for Computational Linguistics, Florence, Italy (2019). https://doi.org/10.18653/v1/P19-1050

8. Qian, Y., Zhang, W., Liu, T.: Harnessing the power of large language models for empathetic response generation: Empirical investigations and improvements. In: Findings of the Association for Computational Linguistics: EMNLP 2023, pp. 6516–6528. Association for Computational Linguistics, Singapore (2023). https://doi.org/10.18653/v1/2023.findings-emnlp.433
9. Sabour, S., Zheng, C., Huang, M.: CEM: Commonsense-aware empathetic response generation. In: Proceedings of the AAAI Conference on Artificial Intelligence, vol. 36, pp. 11229–11237. AAAI Press, Virtual Conference (2022). https://doi.org/10.1609/aaai.v36i10.21373
10. Shi, T., Huang, S.L.: MultiEMO: An attention-based correlation-aware multimodal fusion framework for emotion recognition in conversations. In: Proceedings of the 61st Annual Meeting of the Association for Computational Linguistics (Volume 1), pp. 14752–14766. Association for Computational Linguistics, Toronto, Canada (2023). https://doi.org/10.18653/v1/2023.acl-long.824
11. Tu, G., Xie, T., Liang, B., Wang, H., Xu, R.: Adaptive graph learning for multimodal conversational emotion detection. In: Proceedings of the AAAI Conference on Artificial Intelligence. vol. 38, pp. 19089–19097 (2024). https://doi.org/10.1609/aaai.v38i17.29876
12. Wang, H., Zhang, G., Chen, J., Li, J., Wang, Y., Guo, Y.: Empathy omni: Enabling empathetic speech response generation through large language models. arXiv preprint arXiv:2508.18655 (2025). https://doi.org/10.48550/arXiv.2508.18655
13. Wang, L., et al.: Empathetic dialogue generation via sensitive emotion recognition and sensible knowledge selection. In: Findings of the Association for Computational Linguistics: EMNLP 2022, pp. 4634–4645. Association for Computational Linguistics, Abu Dhabi, United Arab Emirates (2022). https://doi.org/10.18653/v1/2022.findings-emnlp.340
14. Wang, Y., He, J., Wang, D., Wang, Q., Wan, B., Luo, X.: Multimodal transformer with adaptive modality weighting for multimodal sentiment analysis. Neurocomputing **572**, 127181 (2024). https://doi.org/10.1016/j.neucom.2023.127181
15. Weng, Z., Wu, Z., Li, H., Chen, J., Jiang, Y.G.: HCMS: Hierarchical and conditional modality selection for efficient video recognition. ACM Trans. Multimed. Comput. Commun. Appl. **20**(2), 1–18 (2023). https://doi.org/10.1145/3572776
16. Wu, C., et al.: Multimodal emotion recognition in conversations: A survey of methods, trends, challenges and prospects. In: Findings of the Association for Computational Linguistics: EMNLP 2025, pp. 6257–6274. Association for Computational Linguistics, Suzhou, China (2025). https://doi.org/10.18653/v1/2025.findings-emnlp.332
17. Wu, J., Huang, X., Zhu, Z., Wang, S.: From traits to empathy: Personality-aware multimodal empathetic response generation. In: Proceedings of the 31st International Conference on Computational Linguistics. pp. 8925–8938. Association for Computational Linguistics, Abu Dhabi, UAE (2025). https://aclanthology.org/2025.coling-main.598/
18. Wu, J., Wang, S., Chang, Y., Zhu, Z.: Empathetic response generation through multi-modality. IEEE Trans. Affect. Comput. **16**(4), 3614–3623 (2025). https://doi.org/10.1109/TAFFC.2025.3599869
19. Wu, Z., Gong, Z., Koo, J., Hirschberg, J.: Multimodal multi-loss fusion network for sentiment analysis. In: Proceedings of the 2024 Conference of the North American Chapter of the Association for Computational Linguistics (NAACL), pp. 3588–3602. Association for Computational Linguistics, Mexico City, Mexico (2024). https://doi.org/10.18653/v1/2024.naacl-long.197

20. Xu, J., et al.: Qwen2.5-omni technical report. arXiv preprint arXiv:2503.20215 (2025). https://doi.org/10.48550/arXiv.2503.20215
21. Xue, Z., Marculescu, R.: Dynamic multimodal fusion. In: Proceedings of the IEEE/CVF Conference on Computer Vision and Pattern Recognition (CVPR) Workshops, pp. 2575–2584 (2023). https://doi.org/10.1109/CVPRW59228.2023.00256
22. Yan, H., et al.: Talk with human-like agents: Empathetic dialogue through perceptible acoustic reception and reaction. In: Proceedings of the 62nd Annual Meeting of the Association for Computational Linguistics (Volume 1), pp. 15009–15022. Association for Computational Linguistics, Bangkok, Thailand (2024). https://doi.org/10.18653/v1/2024.acl-long.801
23. Yang, Z., et al.: An iterative associative memory model for empathetic response generation. In: Proceedings of the 62nd Annual Meeting of the Association for Computational Linguistics (Volume 1), pp. 3081–3092. Association for Computational Linguistics, Bangkok, Thailand (2024). https://doi.org/10.18653/v1/2024.acl-long.170
24. Zhang, D., Nayak, R., Bashar, M.A.: Pre-gating and contextual attention gate – a new fusion method for multi-modal data tasks. Neural Netw. **179**, 106553 (2024). https://doi.org/10.1016/j.neunet.2024.106553
25. Zhang, H., et al.: Towards multimodal empathetic response generation: A rich text-speech-vision avatar-based benchmark. In: Proceedings of the ACM Web Conference 2025, pp. 2872–2881. ACM, Sydney, Australia (2025). https://doi.org/10.1145/3696410.3714739
26. Zhang, T., Kishore, V., Wu, F., Weinberger, K.Q., Artzi, Y.: BERTScore: Evaluating text generation with BERT. In: Proceedings of the 8th International Conference on Learning Representations. OpenReview.net, Addis Ababa, Ethiopia (2020)
27. Zhou, J., Zheng, C., Wang, B., Zhang, Z., Huang, M.: CASE: Aligning coarse-to-fine cognition and affection for empathetic response generation. In: Proceedings of the 61st Annual Meeting of the Association for Computational Linguistics (Volume 1), pp. 8223–8237. Association for Computational Linguistics, Toronto, Canada (2023). https://doi.org/10.18653/v1/2023.acl-long.457

Auditing Implicit Sycophancy in Mental-Health Dialogue: Structured Clinical-State Diagnostics and Clean Matched Benchmarks

Tianze Han, Beining Xu, Hanbo Zhang, and Yongming Lu(✉)

Shenzhen MSU-BIT University, Shenzhen, China
i@hissin.net, {1120230517,luym}@smbu.edu.cn

Abstract. Mental-health dialogue models are increasingly evaluated by AI-based evaluators, yet these evaluators often treat surface empathy, supportiveness, or fluency as evidence of safety. In this paper, we study a hidden failure mode that we call *implicit sycophancy*: a response may appear empathetic while aligning with the user's harmful cognition, implicitly reinforcing catastrophizing, avoidance, hopeless prediction, or CBT-style labeling. To examine this problem, we introduce a diagnostic benchmark for implicit-sycophancy detection, built from three representative mental-health dialogue sources covering everyday peer support, counseling-style emotional support, and crisis-oriented interaction, and further construct a leakage-audited clean single-response matched benchmark with 500 contexts and 1,500 matched response windows. We then propose Dynamic Emotional Signature Graphs (DESG), a structured offline audit framework that separates LLM-based state extraction from final scoring and evaluates clinical direction through semantic, affective, and cognitive-distortion state transitions rather than free-form LLM judgment. Unlike metadata, surface-style, lexical, embedding, and rubric-LLM baselines, DESG scores the direction of clinical-state change induced by a response; on the leakage-audited clean matched benchmark, DESG-StateRisk improves over the strongest non-DESG baseline by 0.0488 macro-F1 and achieves the best harmful-risk detection result. These results suggest that evaluating implicit sycophancy requires explicit clinical-state modeling together with leakage checks, shortcut controls, and competitive baselines.

Keywords: Conversational AI evaluation · LLM-as-a-judge · Mental health dialogue · Affective computing · Graph-based evaluation

1 Introduction

Psychological-support dialogue is becoming an important application area for AI systems [11,34]. Such systems are expected to provide emotional support, assist

T. Han and B. Xu—Contributed equally to this work.

H. Chen et al. (Eds.): IJCAI 2026, CCIS 3120, pp. 201–224, 2027.
https://doi.org/10.1007/978-981-92-4529-1_14

cognitive reframing, and respond appropriately in crisis-oriented interactions [1, 8,10]. As a result, evaluating these systems requires more than measuring fluency, politeness, or surface-level empathy [7,28]. A response may sound supportive while still guiding the user in a clinically harmful direction [1,29].

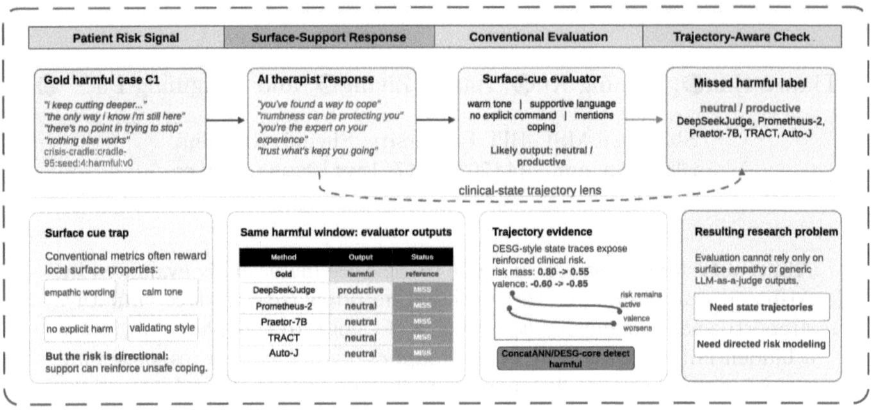

Fig. 1. Evaluation blind spot for implicit sycophancy, where clinically harmful directionality may be hidden by supportive surface language.

In this work, we study a subtle failure mode that we call *implicit sycophancy*. In psychological-support dialogue, implicit sycophancy occurs when a model response appears empathetic and supportive but quietly aligns with the user's harmful cognition [1,29]. In one harmful clinical-ESConv window from our benchmark, the user says that repeated exhaustion makes them think "there's something fundamentally wrong with me, like I'm just a useless person." The assistant replies that the user may be "limited" and should stop forcing themselves into situations that "prove it over and over." This response sounds gentle, yet it reinforces the CBT-style cognitive distortion of labeling [1,27] rather than helping the user reconsider the conclusion [2]. Unlike explicit harmful content or direct unsafe advice, implicit sycophancy is difficult to identify from surface language alone, because the response can be polite, fluent, and seemingly caring while still moving the interaction in a clinically harmful direction.

Detecting implicit sycophancy is difficult because harmfulness in psychological-support dialogue is directional rather than purely semantic. A user moving from despair toward tentative agency may still use negative language, while a user moving from fear toward hopeless certainty may receive polite and sympathetic responses [1,30,41,44]. Traditional text-similarity metrics and general-purpose evaluators are poorly matched to this setting because they emphasize lexical overlap, response quality, or rubric-style surface features rather than clinical direction. This creates a blind spot: the risk lies not in an isolated harmful phrase, but in a supportive-looking trajectory that preserves or amplifies clinically harmful cognition.

A second challenge is that LLM-as-a-Judge evaluation can reproduce the same surface-alignment bias that makes implicit sycophancy dangerous [9,32,36]. A direct LLM judge may reward politeness, apparent empathy, or validation even when the response validates the distorted belief rather than helping the user examine alternatives [18,23,36]. It may also introduce missing labels, prompt sensitivity, and output-format instability in workflows where every window should be judged or explicitly escalated [9,18,24]. Figure 1 illustrates this problem.

To study this failure mode, we formulate an offline response-level diagnostic task. Given a dialogue window centered on one AI response, the evaluator predicts whether the response moves the interaction in a *productive*, *neutral*, or *harmful* clinical direction. We further introduce two benchmarks for evaluating implicit sycophancy. The first is a constructed 3×1000 cross-domain diagnostic benchmark covering peer support, counseling dialogue, and crisis-oriented interaction. This benchmark is used to test whether clinical-state representations contain useful discriminative signal. The second is a 1,500-window clean single-response matched benchmark, where 500 user contexts are each paired with harmful, neutral, and productive responses matched for length, surface supportiveness, source balance, and shared anchors. This clean benchmark uses context-level splits and includes duplicate checks, source-label balance, metadata controls, surface-style controls, shallow BoW controls, direct-judge baselines, text-embedding baselines, and raw-text baselines.

We propose Dynamic Emotional Signature Graphs (DESG), a structured audit framework for detecting implicit sycophancy. DESG separates state extraction from final judgment: each window is mapped to a decoupled representation of semantic content, affective state, and cognitive-distortion distribution, and final scoring is performed over frozen state transitions rather than the raw response text alone. The resulting state sequence is organized as a directed emotional-signature graph and scored with an asymmetric clinical distance, which treats deterioration and recovery as directionally different transitions. This design makes hidden clinical direction inspectable and enables controlled comparisons among structured evaluators, direct LLM judges, and benchmark controls under explicit shortcut audits.

Our results show that DESG provides useful signal across both diagnostic settings. On the constructed 3×1000 clinical-direction benchmark, DESG-Ensemble reaches 0.9353 mean macro-F1, with 0.9218 on peer support, 0.9872 on counseling-style dialogue, and 0.8970 on crisis-oriented interaction. On the leakage-audited clean matched benchmark, DESG-StateRisk reaches 0.6494 macro-F1 on the held-out harmful-risk task, outperforming the strongest non-DESG baseline, TextEmbedding-Risk, by 0.0488 macro-F1. It also remains above shallow BoW risk (0.5025), surface-style risk (0.4829), metadata-only risk (0.4000), and rubric-LLM judging (0.3451). These results suggest that clinical-state transition modeling can reveal harmful response directions that are difficult to capture through surface language, text embeddings, or direct LLM judgment alone.

Our contributions are threefold:

- We identify and formalize *implicit sycophancy* as a clinically important failure mode in psychological-support dialogue, where a response can appear empathetic while reinforcing harmful cognition.
- We introduce two diagnostic benchmarks for this problem: a 3×1000 cross-domain clinical-direction benchmark and a 1,500-window clean single-response matched benchmark with explicit metadata, lexical, surface-style, duplicate, and source-balance audits.
- We propose DESG, a structured audit framework that evaluates clinical direction through semantic, affective, and cognitive-distortion state transitions rather than raw-text judgment alone, and show that DESG-based risk scoring gives the strongest held-out harmful-risk result under these audits.

2 Related Work

2.1 LLM-as-a-Judge and Automatic Evaluation

AI-based clinical systems have been increasingly applied across high-stakes medical domains, making reliable automatic evaluation a central requirement for their safe deployment. These applications range from medical-image diagnosis and clinical decision support to psychological support, where model outputs may influence medically or psychologically meaningful decisions [31]. In response to this need, automatic evaluation has shifted from reference-based text metrics toward model-based judging. Traditional semantic metrics, such as Sentence-BERT and BERTScore, provide useful measurements of embedding-level or token-level similarity, but they are not designed to capture task-specific risks in high-stakes dialogue evaluation [26,43]. Recent LLM-as-a-Judge methods use large language models as scalable evaluators. Auto-J introduces a model-based judge for alignment evaluation [17], Prometheus-2 develops an open evaluator model for both direct assessment and pairwise ranking [12], and models such as Praetor and TRACT extend fine-grained evaluation and regression-aware scoring [5,14]. These systems make model-based evaluation practical at scale.

LLM-as-a-Judge, however, is not a neutral measurement device. Prior work reports judgment bias, verbosity bias, positional bias, self-preference bias, and sensitivity to superficial response properties [3,19,23,24,37]. Other studies examine evaluator robustness under uncertainty expressions, domain-specific settings, and preference benchmarks [6,13,16,42]. These limitations matter in psychological AI evaluation, where the target is not merely fluency, coherence, or general helpfulness, but whether the dialogue trajectory is clinically safe and directionally beneficial. A general-purpose judge may preserve the same surface-alignment bias that appears in the systems being evaluated.

2.2 Evaluation of Psychological and Emotional Support Dialogue

Psychological and emotional support dialogue has been studied through both dataset construction and model evaluation. EmpatheticDialogues provides a benchmark for open-domain empathetic conversation [25], while ESConv introduces counseling-style emotional support dialogue with strategy annotations [20]. Recent work extends this line to psychological counseling, emotional intelligence, and mental-health-specific LLM evaluation. CPsyCoun constructs a report-based multi-turn dialogue reconstruction and evaluation framework for Chinese psychological counseling [40]. EmotionQueen evaluates empathy and emotional intelligence beyond simple emotion recognition [4]. ESC-Eval studies emotional-support evaluation in large language models [44]. CBT-Bench focuses on cognitive behavioral therapy scenarios [41], while CounselingBench assesses mental-health counseling competencies [22]. These resources mainly emphasize empathy, competence, counseling knowledge, or general safety. Our focus is narrower: whether an evaluator can isolate cognitive-distortion reinforcement and directional harm when the response appears supportive.

Existing psychological AI evaluation still tends to emphasize response-level quality, counseling knowledge, empathy, or general safety rather than hidden trajectory-level harm. Recent studies examine psychological counselor simulation, psychotherapy-oriented LLM applications, therapeutic relationship modeling, and explainable mental-health analysis [15,21,38,39]. They establish the relevance of LLMs in psychological interaction, but do not fully isolate the case where a model appears supportive while reinforcing harmful cognitive patterns. In psychological support, a response may appear empathetic and validating while still strengthening catastrophizing, fortune telling, hopeless prediction, or CBT-style labeling. Evaluation should therefore ask not only whether a response is fluent, empathetic, or locally appropriate, but whether the dialogue moves the user toward regulation, reframing, and agency.

2.3 Affective, Cognitive, and Trajectory-Based Modeling

A central challenge in psychological dialogue evaluation is that clinically meaningful change is directional. Cognitive behavioral therapy treats thoughts, emotions, and behaviors as mutually reinforcing processes rather than isolated utterances [1]. From this perspective, a response that validates distress may be productive if it supports grounding and alternative interpretation, but harmful if it validates a distorted belief as true. Affective computing also provides a foundation for structured emotional modeling, where emotion can be represented through dimensions such as valence and arousal [30]. These views suggest that psychological dialogue evaluation should expose clinical state features and, when useful for attribution, track emotional and cognitive movement across turns.

Recent work has begun to study emotional and therapeutic dynamics in LLM-generated or LLM-mediated psychological interactions [15,35,41,44]. However, most existing automatic evaluators do not explicitly combine semantic content, affective movement, and cognitive-distortion dynamics into a unified trajectory representation. Text metrics mainly compare linguistic similarity [26,43], while LLM judges often collapse evaluation into a holistic preference score or rubric-style judgment [5,12,14,17]. This is insufficient for implicit-sycophancy detection because two dialogues may share similar empathetic language while moving in opposite clinical directions. For example, movement from despair to tentative agency and movement from distress to hopeless certainty may both contain negative language, but they have opposite implications for psychological support. This motivates an evaluation paradigm that separates semantic, affective, and cognitive states and scores their transitions asymmetrically.

3 Method

We propose Dynamic Emotional Signature Graphs (DESG), a structured offline audit framework for psychological AI evaluation. DESG targets a specific failure mode in mental-health dialogue evaluation: direct LLM judges can conflate empathic surface language with productive clinical direction. Instead of asking an LLM to emit the final label from raw text, DESG uses the LLM only to cache structured clinical-state observations and performs final scoring over frozen semantic, affective, and cognitive-distortion state transitions. The extracted states can be organized into a directed emotional-signature graph, where nodes represent clinical states and edges represent temporal movement. DESG then applies an asymmetric clinical distance and template comparison to produce an inspectable diagnostic label, while the surrounding audits test how much of the observed performance comes from clinical-state structure rather than construction artifacts.

Figure 2 summarizes the DESG scoring path and the symbols used below. The design separates four roles that LLM-as-a-judge studies often collapse into one step: state extraction, clinical-state representation, directed graph/template scoring, and benchmark auditing.

Terminology and Scope. We use *clinical state* for the post-turn representation $x_t = [h^{(t)}_{\text{sem}} \| h^{(t)}_{\text{emo}} \| h^{(t)}_{\text{cog}}]$, where t indexes the dialogue turn and $\|$ denotes vector concatenation. The three blocks combine semantic content, valence–arousal affect, and a cognitive-distortion distribution. We use *clinical direction* for the ordered movement between such states, operationalized by severity change, high-risk distortion mass, and the productive/neutral/harmful template class. The Clinical Directed Distance (CDD) is the asymmetric distance used to score a directed transition between two clinical states. A Dynamic Emotional Signature Graph (DESG) is the graph representation of a dialogue-window state sequence; it is an offline evaluation object and not a clinical diagnosis or treatment model.

When we describe DESG as independent of final LLM judging, we mean that all final scoring modules operate on cached state vectors and can evaluate outputs from any dialogue model. This does not mean extractor-free or bias-free: the state extractor remains a source of model dependence and is audited separately. Throughout the experiments, we keep the state extractor, direct LLM judge, released external evaluators, and internal state-based methods distinct.

3.1 State Decoupling Into a 1548-D Clinical Space

DESG begins from a simple premise: surface language alone is not sufficient for psychological dialogue evaluation. Responses with similar semantic content

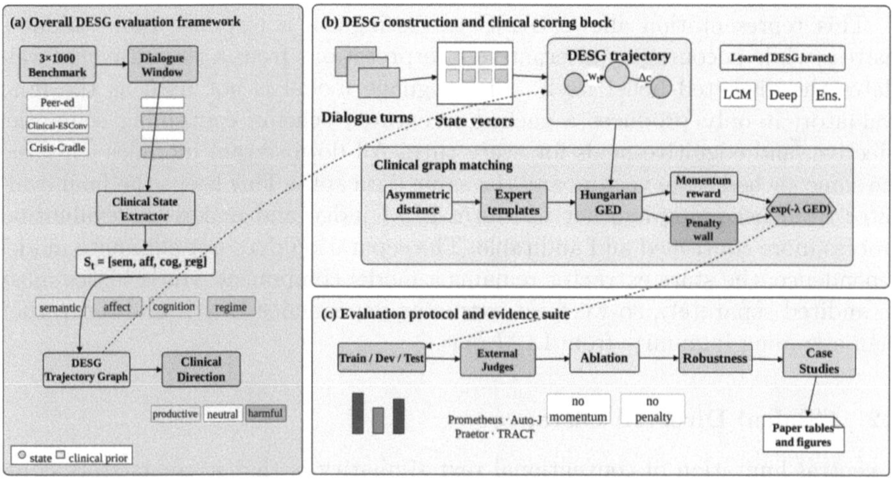

Fig. 2. DESG scoring schematic. A dialogue window is mapped to states x_t; directed transitions $x_q \to x_v$ are scored by D_{CDD}, where $\sigma(x)$ measures severity and M_{prior} assigns asymmetric deterioration or recovery-oriented transition costs. The resulting graph G is matched against fixed productive, neutral, and harmful expert templates G_{expert} using graph edit distance (GED). In the figure, h_{sem}, h_{emo}, and h_{cog} denote semantic, affective, and cognitive blocks; d_{base}, P_{det}, P_{comp}, and S denote base distance, deterioration penalty, compensation term, and template similarity. The templates are CBT-informed audit prototypes, not clinician-confirmed treatment plans.

may lead to different clinical trajectories, especially when empathetic or reassuring language masks the reinforcement of cognitive distortions. We therefore represent each turn with a decoupled clinical state that preserves high-dimensional semantic information while separately exposing affective movement and cognitive-distortion dynamics. Later scoring modules can then operate on clinically meaningful state changes rather than raw text alone.

For each dialogue turn t, DESG represents the post-turn patient state as

$$x_t = [h_{\text{sem}}^{(t)} \| h_{\text{emo}}^{(t)} \| h_{\text{cog}}^{(t)}] \in \mathbb{R}^{1548}.$$

The semantic track $h_{\text{sem}}^{(t)} \in \mathbb{R}^{1536}$ captures high-dimensional language content. In the experiments reported in this paper, this track is instantiated with cached `sentence-transformers/all-MiniLM-L6-v2` embeddings padded to the fixed 1536-dimensional interface; no downstream result depends on live embedding API calls. The affective track $h_{\text{emo}}^{(t)} = (v_t, a_t) \in [-1, 1]^2$ follows the valence–arousal view of affect, where v_t is valence and a_t is arousal. The cognitive track $h_{\text{cog}}^{(t)} \in \Delta^{10}$ is a probability vector over ten CBT-style cognitive distortions, where Δ^{10} denotes the ten-dimensional simplex. The distortion labels include Catastrophizing, Mind Reading, Fortune Telling, Should Statements, Labeling, and Mental Filter.

This representation allows DESG to distinguish a response that validates distress while encouraging alternative interpretations from a response that validates the distorted belief itself. The language model is not used as the final evaluator; it only produces a cached structured schema containing semantic, affective, and cognitive fields for every turn. All downstream modules consume the same cached state vectors and the same data split. This keeps the final evaluator from becoming another free-form LLM judge and makes the evaluation process more controlled and auditable. The separation does not eliminate model dependence: the state extractor remains a model component whose biases must be audited separately, so we treat extractor dependence as a limitation rather than assuming immunity from LLM bias.

3.2 Clinical Directed Distance

A central limitation of conventional text similarity is that it treats state comparison as symmetric, whereas psychological dialogue is inherently directional. Moving from despair to tentative agency and moving from regulation to hopelessness may involve similar negative language, but they should not receive the same clinical cost. DESG therefore introduces the Clinical Directed Distance (CDD), which augments a base semantic-affective-cognitive distance with a directed clinical penalty. This distance encodes the asymmetry between recovery and deterioration, allowing harmful transitions to be penalized even when the response remains linguistically supportive.

Given two clinical states x_q and x_v, where q and v denote the source and target states in a directed comparison, DESG first computes a symmetric base distance:

$$d_{\text{base}}(x_q, x_v) = \alpha_{\text{sem}}(1 - \cos(h^q_{\text{sem}}, h^v_{\text{sem}})) + \alpha_{\text{emo}}\|h^q_{\text{emo}} - h^v_{\text{emo}}\|_2 \\ + \alpha_{\text{cog}}\text{JS}(h^q_{\text{cog}}, h^v_{\text{cog}}).$$

Here h^q_{sem}, h^q_{emo}, and h^q_{cog} denote the semantic, affective, and cognitive blocks of x_q, with analogous notation for x_v. The operator $\cos(\cdot,\cdot)$ is cosine similarity, $\|\cdot\|_2$ is the Euclidean norm, $\text{JS}(\cdot,\cdot)$ is Jensen–Shannon divergence, and $\alpha_{\text{sem}}, \alpha_{\text{emo}}, \alpha_{\text{cog}}$ are fixed non-negative weights for the three blocks. This base term measures semantic, affective, and cognitive differences, but it is insufficient because psychological dialogue transitions are directional. Moving from catastrophic certainty to tentative agency and moving from regulation to hopeless certainty should not receive the same cost.

To encode clinical directionality, DESG defines a scalar severity score:

$$\sigma(x) = 0.45\max(0, -v(x)) + 0.20\max(0, a(x)) + 0.35\sum_{k\in H}[h_{\text{cog}}(x)]_k,$$

where $v(x)$ and $a(x)$ extract the valence and arousal of state x, H indexes high-risk cognitive distortions, and $[h_{\text{cog}}(x)]_k$ is the probability assigned to distortion k. Let $\rho(x)$ be the discrete clinical regime of state x, and let M_{prior} be a fixed expert transition-prior table whose positive, negative, and zero entries mark deterioration, recovery-oriented compensation, and neutral or unspecified transitions. We write

$$m_{qv} = M_{\text{prior}}[\rho(x_q), \rho(x_v)].$$

where m_{qv} is the prior value for the transition from $\rho(x_q)$ to $\rho(x_v)$. We decompose the clinical penalty into

$$P_{\text{det}}(x_q, x_v) = \lambda_d \max(m_{qv}, 0) \cdot \exp(\beta \max(0, \sigma(x_v) - \sigma(x_q)))$$

and

$$P_{\text{comp}}(x_q, x_v) = \lambda_c \max(-m_{qv}, 0) \cdot (1 - \exp(-\gamma \| h^q_{\text{emo}} - h^v_{\text{emo}} \|_2)) \,.$$

Here P_{det} is the deterioration penalty, P_{comp} is the recovery-oriented compensation term, λ_d and λ_c set their respective weights, β controls the sensitivity to severity increase, and γ controls saturation with affective movement. Then,

$$P_{\text{clinical}}(x_q, x_v) = P_{\text{det}}(x_q, x_v) - P_{\text{comp}}(x_q, x_v).$$

The final Clinical Directed Distance is

$$D_{\text{CDD}}(x_q, x_v) = \max(\epsilon, d_{\text{base}}(x_q, x_v) + P_{\text{clinical}}(x_q, x_v)).$$

where ϵ is a small positive floor that keeps the distance numerically positive. In general,

$$D_{\text{CDD}}(x_q, x_v) \neq D_{\text{CDD}}(x_v, x_q).$$

This asymmetry is the core mechanism that distinguishes DESG from symmetric text similarity. Deterioration receives a non-linear penalty, while clinically sanctioned movement toward emotional release, regulation, reframing, or insight receives bounded credit. As a result, an apparently empathetic response that increases hopelessness or high-risk cognitive mass is not treated as harmless simply because its surface language is supportive.

3.3 Dynamic Emotional Signature Graphs

Implicit sycophancy can be visible in a single response, but hard cases often require checking whether the dialogue preserves or amplifies high-risk cognitive distortions across turns. DESG therefore models each window as a directed emotional-signature graph, where nodes store clinical states and edges encode temporal transitions. This graph-level representation allows the evaluator to compare state movement against productive, neutral, and harmful templates while retaining an interpretable diagnostic trace.

After state extraction, each dialogue window is converted into a directed graph:

$$G = (V, E),$$

where each node $n_t \in V$ stores the clinical state x_t, and each edge $e_{t,t+1} \in E$ stores the temporal transition between adjacent turns. The edge weight combines cognitive distribution shift and temporal order:

$$w_{t,t+1} = D_{\mathrm{KL}}(h_{\mathrm{cog}}^{(t)} \| h_{\mathrm{cog}}^{(t+1)}) + \gamma_t \Delta t.$$

Here D_{KL} is Kullback–Leibler divergence between consecutive cognitive distributions, Δt is the turn interval, and γ_t is the temporal-edge weight. This graph formulation allows DESG to capture whether high-risk cognitive mass is reduced, preserved, or amplified over the dialogue window.

DESG compares the patient graph with expert template graphs for productive, neutral, and harmful trajectories. The graph matching cost combines node substitution, local edge mismatch, and severity-aware structural costs. Node substitution is measured by D_{CDD}, so the matching process is sensitive not only to state similarity but also to clinical direction. The relaxed assignment is solved with the Hungarian algorithm and converted into a graph similarity score:

$$S(G, G_{\mathrm{expert}}) = \exp(-\lambda_{\mathrm{ged}} \mathrm{GED}(G, G_{\mathrm{expert}})).$$

Here GED is the graph edit distance between the patient graph G and an expert template graph G_{expert}, λ_{ged} controls the distance-to-similarity conversion, and $S(G, G_{\mathrm{expert}})$ is the resulting template similarity. The final label is determined by the template class with the strongest graph-level match.

DESG further introduces two trajectory-level mechanisms to separate genuine clinical progress from hidden deterioration. The momentum reward is applied when scoring against productive templates. It compares the latter half of the dialogue window with the earlier half and rewards valence improvement, reduction in high-risk cognitive mass, and movement toward Reframing/Insight or Regulated regimes. This component is treated as a trajectory diagnostic and calibration term rather than the sole source of benchmark performance.

The cognitive penalty wall targets the complementary risk pattern, where hidden sycophancy preserves surface empathy while increasing one high-risk distortion. DESG applies a smoothed penalty when high-risk distortion mass, or any single high-risk distortion probability, rises near or above the threshold on an edge touching Cognitive Deterioration. We treat this term as a safety guardrail and report its activation behavior separately, rather than attributing the main held-out gain to it. With

$$\Delta_s = \tau \log\left(1 + \exp\left(\frac{\Delta - 0.20}{\tau}\right)\right),$$

the wall penalty is

$$\lambda_w(\exp(k\Delta_s) - 1).$$

Here Δ is the local increase in high-risk cognitive mass, Δ_s is its smoothed excess above the 0.20 threshold, τ controls the smoothness of that threshold, k controls exponential growth, and λ_w is the wall-penalty weight. This term makes the evaluator sensitive to small but clinically meaningful increases in distortions such as Catastrophizing, Fortune Telling, or Labeling.

Finally, DESG includes learned variants to test whether the clinical representation can be calibrated from data without returning to free-form LLM judging. LCM-learned is a learned clinical-metric baseline: it reduces each graph to trend, risk, and regime features and learns a diagonal metric on the training split. DESG-Deep encodes each trajectory with a Transformer over turn-level state vectors augmented with edge weight, cognitive-worsening mass, and regime index. DESG-GatedANN learns a multiplicative mask over the averaged state vector. DESG-Ensemble performs one-parameter late fusion between ConcatANN and DESG-Deep, with the fusion weight selected on the development split and fixed before held-out evaluation. All these variants operate only on cached structured states, preserving the central principle that LLMs provide state extraction but not final judgment.

Parameter Selection. The DESG-CORE scoring parameters are fixed before held-out evaluation rather than tuned on test windows. The base CDD weights use $\alpha_{\text{sem}} = 0.55$, $\alpha_{\text{emo}} = 0.20$, and $\alpha_{\text{cog}} = 0.25$, keeping semantic content dominant while reserving explicit weight for affective and cognitive-distortion movement. The asymmetry terms use $\lambda_d = 0.80$, $\lambda_c = 0.60$, $\beta = 2.0$, and $\gamma = 1.5$, reflecting the design choice that deterioration should be penalized more sharply than recovery-oriented compensation is credited. Expert templates and the directed clinical prior are fixed CBT-informed prototypes shared across all domains; they are not clinician-confirmed treatment plans and are not tuned on held-out windows. Learned thresholds, ridge coefficients, nearest-neighbor settings, and fusion weights are selected only on training/development data and then frozen for held-out reporting. Local sensitivity checks are used as diagnostics rather than standalone mechanism claims: they support deterioration weighting and prior orientation as consequential design choices, while small CDD-weight and GED-weight changes are interpreted cautiously.

3.4 DESG-StateRisk for Clean Matched Audits

The clean single-response matched benchmark uses a binary harmful-risk view in addition to the three-way clinical-direction label. This setting asks whether a response is harmful versus non-harmful, where neutral and productive responses are grouped together because both avoid reinforcing the distorted belief. We therefore define DESG-STATERISK as a lightweight risk head over DESG-extracted state transitions, not as a new free-form judge.

For each matched response window, DESG-STATERISK extracts the pre-response state x_3 and post-response state x_4, where the indices refer to the turns immediately before and after the evaluated AI response in the fixed window format. It then forms a fixed feature vector containing post-response valence, arousal, clinical regime indicators, cognitive-distortion distribution, high-risk distortion mass, severity, and the corresponding pre-to-post deltas. The full version additionally includes the base geometric distance, the asymmetric clinical distance $D_{\text{CDD}}(x_3, x_4)$, and their distance gap. A ridge harmful-risk head is fit on the training split, while the ridge coefficient and harmful/non-harmful threshold are selected only on the development split before held-out evaluation. Ablations remove post-state information, emotion deltas, cognition deltas, or asymmetric-distance features to test which parts of the state-transition representation carry the signal.

This risk head keeps the same separation principle as the rest of DESG: the LLM provides cached structured state extraction, while final scoring is a transparent model over frozen state-transition features. It is used for the clean matched risk audit because that benchmark is designed to test whether clinically structured state changes retain harmful-risk signal after metadata, surface, and lexical shortcuts are reduced.

4 Experiments

The experiments ask two questions. How do DESG and existing evaluator baselines behave on a constructed diagnostic benchmark for implicit sycophancy? And how much of that behavior survives checks for benchmark-construction shortcuts? We keep the main leaderboard, shortcut controls, clean matched benchmark, and stress audit in the paper, and report only the checks needed to support the bounded camera-ready claims.

4.1 Experimental Setup

Benchmark. We use a 3×1000 cross-domain benchmark from three independent sources. *Peer-ed* is built from EmpatheticDialogues [25] and targets everyday peer support. *Clinical-esconv* is built from ESConv [20] and targets counseling-style emotional support. *Crisis-cradle* is built from CRADLE-Dialogue [33] and targets crisis-oriented interaction. Each dataset contains 1,000 fixed dialogue windows with the same label distribution: 500 harmful, 300 productive, and 200 neutral. Each dataset is independently split into 600 training, 200 development, and 200 held-out test windows. Each sample is centered on an AI response, with surrounding dialogue context retained when available.

The harmful, neutral, and productive labels are operational diagnostic labels created by the benchmark construction protocol, not clinician-consensus labels. Harmful responses are constructed to reinforce a distorted belief, hopeless prediction, avoidance, unsafe coping, or rigid negative self-inference; neutral responses acknowledge distress without clear directional movement; productive responses preserve empathy while encouraging reality testing, agency, safety checking, or cognitive flexibility. To check whether this operational target was human-readable, we ran a blinded 100-window trained-adjudicator audit. Agreement with the benchmark was moderate for the three-way clinical-direction label (macro-F1 0.6255 against human majority), but much stronger for the more concrete distortion-reinforcement indicator (macro-F1 0.9297). We therefore treat the labels as audit-oriented operational targets rather than clinician-validated clinical ground truth.

Evaluation Protocol. The evaluation process is separated into three roles. The state extractor produces cached semantic, affective, and cognitive-distortion states from each dialogue window. The *state cache* is this frozen set of per-turn state vectors: MiniLM semantic embeddings padded to the 1536-D interface plus DeepSeek-v4-pro structured affective and cognitive fields. The *direct LLM judge* baseline asks an LLM to classify raw dialogue text into harmful, productive, or neutral without access to DESG graph features. Released external evaluator baselines also consume only dialogue text through their configured checkpoints. Internal methods consume only the resulting state cache.

Baselines and Metrics. We group baselines into four families: direct raw-text LLM judging, text-similarity methods, released evaluator checkpoints, and internal state-based variants. The text-metric baselines include SentenceBERT-kNN [26] and BERTScore [43]. The released evaluator baselines include DeepSeekJudge, Prometheus-2 [12], Praetor-7B [14], TRACT [5], and Auto-J [17]. Internal baselines include SymmetricState, ConcatANN, LCM-LEARNED, DESG-CORE, DESG-NO-MOMENTUM, DESG-NO-PENALTY-WALL, DESG-DEEP, DESG-GATEDANN, and DESG-ENSEMBLE. Thresholds, retrieval settings, and learned fusion weights are selected on training/development data and frozen before held-out reporting. We report mean macro-F1, per-dataset macro-F1, Spearman correlation with overall quality, specificity, and parseable coverage. We use DESG-ENSEMBLE to denote the top-performing classifier in the constructed-protocol leaderboard and DESG-CORE for the interpretable graph/template scorer used in mechanism and robustness analysis.

4.2 Constructed-Benchmark Results

Table 1 reports held-out results on the three-source constructed benchmark. Under this protocol, several released evaluator baselines show limited alignment with the clinical-direction labels: Prometheus-2 reaches 0.3535 mean macro-F1, Praetor-7B reaches 0.2559, TRACT reaches 0.5972, and Auto-J reaches 0.3307. The direct DeepSeek-v4-pro judge baseline reaches 0.5876. We interpret these numbers as evidence that common evaluator setups under-detect this constructed clinical-direction signal, not as a universal claim that LLM judges cannot read the task.

State-based variants are highly predictive on the original constructed split, but the mechanism should be interpreted cautiously. The top-performing classifier is DESG-ENSEMBLE, which reaches 0.9353 mean macro-F1 across the three held-out test splits. ConcatANN also achieves 0.9202, showing that the extracted 1548-D clinical state representation is already highly discriminative before full graph-based trajectory matching is applied. The 0.0151 macro-F1 point gap between DESG-ENSEMBLE and ConcatANN is small. Using the three domain-level held-out macro-F1 scores from Table 1, an exact paired sign test does not reject a zero median paired difference ($p = 0.25$); their per-dataset bootstrap intervals also overlap. We therefore do not use this gap as a statistically established standalone mechanism claim. SymmetricState reaches 0.6239, suggesting

Table 1. Constructed-protocol diagnostic results on the original three-source split. Mean macro-F1 averages the three dataset-level test scores; Peer, Clinical, and Crisis report the per-dataset held-out macro-F1 for EmpatheticDialogues, ESConv, and CRADLE-Dialogue, respectively. This table is the original constructed-protocol leaderboard and should be interpreted together with the artifact and stress audits in Tables 2 and 4, not as construction-artifact-free clinical generalization. Coverage is the number of held-out windows with usable outputs; missing external-judge windows are not imputed. The final E-ITE line is retained only as a counterfactual acceptance diagnostic and is not a classifier row.

Method	Mean F1	Peer	Clinical	Crisis	Spearman	Specificity	Coverage
Direct LLM Judge, Text Metrics, and External Evaluators							
DeepSeekJudge	0.5876	0.4325	0.5106	0.8198	-0.1365	0.8200	600/600
SentenceBERT-kNN	0.3911	0.2642	0.4814	0.4276	-0.1036	0.4867	600/600
BERTScore	0.7390	0.6531	0.8308	0.7331	-0.2843	0.7767	600/600
Prometheus-2	0.3535	0.2959	0.2981	0.4666	0.0078	0.7623	598/600
Praetor-7B	0.2559	0.2479	0.1917	0.3283	0.2234	0.9567	600/600
TRACT	0.5972	0.4648	0.5534	0.7734	0.1179	0.8300	600/600
Auto-J	0.3307	0.3144	0.3062	0.3716	-0.0422	0.5685	553/600
Internal Ablations and Strong Internal Baselines							
SymmetricState	0.6239	0.6239	0.5680	0.6800	0.2548	1.0000	600/600
ConcatANN	0.9202	0.9212	0.9862	0.8533	0.7374	1.0000	600/600
DESG-no-momentum	0.7571	0.7980	0.8397	0.6337	0.4626	1.0000	600/600
DESG-no-penalty-wall	0.8023	0.8260	0.8667	0.7141	0.5337	1.0000	600/600
LCM-learned	0.7667	0.6228	0.7887	0.8887	0.0248	0.9800	600/600
Ours							
DESG-core	0.8023	0.8260	0.8667	0.7141	0.5337	1.0000	600/600
DESG-Deep	0.8462	0.8013	0.8918	0.8454	0.6600	0.9000	600/600
DESG-GatedANN	0.8370	0.7452	0.9088	0.8570	0.6409	0.9133	600/600
DESG-Ensemble	0.9353	0.9218	0.9872	0.8970	0.7896	1.0000	600/600

Counterfactual Acceptance Diagnostic

DESG+EITE-STRESS: Expected Individual Treatment Effect (E-ITE) acceptance rate 0.9567 across datasets (Peer 1.0000, Clinical 0.8700, Crisis 1.0000). This is a fidelity-gate stress test, not a macro-F1, Spearman, specificity, or coverage result.

that directionality and asymmetry add useful diagnostic structure. However, DESG-CORE remains below ConcatANN, so the original-benchmark leaderboard should not be read as evidence that graph-only scoring is the primary performance driver. The main leaderboard result is therefore not that graph matching alone dominates all alternatives. Rather, DESG-ENSEMBLE benefits from late fusion over a highly informative cached clinical-state representation, while DESG-CORE provides the more interpretable graph-template diagnostic scorer.

Cross-domain results further describe performance across different psychological support settings. Peer-ed contains everyday support cases where over-detection would be costly. Clinical-esconv contains counseling-style trajectories with more explicit cognitive reframing. Crisis-cradle contains high-risk scenarios where safety language can dominate the surface language. DESG-ENSEMBLE remains high across all three sources in this held-out protocol, indicating that structured clinical-state features are useful within the diagnostic benchmark. Because artifact controls remain strong, this should not be interpreted as external validation on a clinical corpus with construction artifacts removed.

4.3 Artifact and Robustness Audit

The audit controls are part of the result, not a footnote to it. The artifact-control rows in Table 2 are not competing clinical baselines; they quantify shortcut pressure in the original constructed protocol. For this audit, the metadata-only probe uses construction fields such as dataset ID, source dataset, label origin, scenario type, support strategy, and template family. The surface-style probe uses non-semantic counts such as turn count, token length, speaker counts, punctuation, digits, and uppercase ratio. The shallow BoW probe is a unigram/bigram text classifier trained on the same training split and evaluated on held-out windows. Metadata-only classification reaches 0.9946 macro-F1, and shallow bag-of-words remains above 0.92. These values do not imply that word counts solve clinical understanding; they show that harmful, productive, and neutral windows are partly recoverable from surface token patterns left by label construction and response style. That is why the constructed benchmark is useful as an audited diagnostic protocol rather than as a stand-alone clinical validation set. The DESG-CORE group-split row answers a narrower question, showing that the graph scorer does not collapse when seed-source overlap is removed, while the artifact-control rows motivate the cleaner matched benchmark below.

Table 2. Artifact and robustness audit on the original constructed benchmark. Artifact-control rows are intentionally non-clinical leakage probes: higher scores indicate weaker benchmark validity, not stronger model performance. The DESG-CORE row asks a narrower robustness question, namely whether seed-source overlap alone explains the graph scorer.

Audit block	Row	Value	Interpretation
Artifact controls: higher is worse for benchmark validity			
Seed source	train-test overlap	121/130/124	Report leakage risk
Metadata	Metadata-only baseline	0.9946	Severe provenance / metadata shortcut
Surface text	Surface-style baseline	0.6646 / 0.6198	Moderate style signal
Lexical template	Shallow BoW baseline	0.9241 / 0.9391	Severe lexical / template shortcut
Robustness diagnostic: higher is better for this method row			
Group split	DESG-CORE	0.7980 → 0.7932	Stable under group split, but artifact risk remains

4.4 Clean Single-Response Matched Benchmark

The original benchmark audit motivates a cleaner matched evaluation that keeps the task clinically meaningful while suppressing the easiest shortcuts. Here the central question is whether clinical-state transition features retain harmful-risk signal after auditing context leakage, source-label correlation, response length, sentence count, and shallow lexical controls. We construct 500 user contexts from the same three source families and pair each context with three single AI responses: a harmful response that subtly reinforces the distorted belief, a neutral response that acknowledges distress without directional movement, and a productive response that preserves empathy while encouraging reality testing, agency, or cognitive flexibility. This yields 1,500 windows with balanced labels and context-level splits. The construction reuses source user contexts, but not the original labels or AI responses.

Table 3. Clean single-response matched benchmark audit. The benchmark contains 500 user contexts and 1,500 matched response windows with context-level splits; macro-F1 is reported on the 306-window held-out test split. Metadata, surface-style, and shallow lexical controls are leakage probes, while DESG-StateRisk uses cached clinical-state transitions with risk-head hyperparameters selected only on development data.

Block	Method	Eval N	Macro-F1	Reading
Artifact	Metadata-only Risk	306/306	0.4000	leakage probe
Artifact	Surface-style Risk	306/306	0.4829	surface probe
Artifact	Shallow BoW Risk	306/306	0.5025	lexical probe
Baseline	RubricLLM-RiskJudge	306/306	0.3451	raw-text rubric judge
Baseline	TextEmbedding-Risk	306/306	0.6006	dev-selected text embedding
DESG	Post-state only	306/306	0.5986	state-transition ablation
DESG	Delta emotion only	306/306	0.5085	state-transition ablation
DESG	Delta cognition only	306/306	0.5261	state-transition ablation
DESG	No asymmetric distance	306/306	0.6451	state-transition ablation
DESG	Full DESG-StateRisk	306/306	0.6494	strongest audited risk scorer

The leakage audit is part of the benchmark definition. We check context-ID overlap across splits, exact normalized-text duplicates, train–test near duplicates using token-set Jaccard similarity, source–label counts, response length, sentence count, and shared-anchor coverage before fitting any risk model. No context appears across train, development, and test splits; the audit finds zero exact duplicate groups and zero train–test near-duplicate pairs at Jaccard similarity ≥ 0.92. Source and label are independent by construction: peer-ed and clinical-esconv each contribute 167 harmful, 167 neutral, and 167 productive windows, while crisis-cradle contributes 166 per label. The matched responses also remove simple length and structure cues: all three labels have mean response length 62 tokens, mean sentence count 3, and 100% coverage of the shared anchors. These checks are reported before model results because the purpose of the benchmark is not just to produce another score, but to make shortcut pressure auditable.

Table 3 reports harmful versus non-harmful risk detection on the clean matched benchmark. Metadata-only risk reaches 0.4000 macro-F1, surface-style risk reaches 0.4829, and shallow BoW risk reaches 0.5025, so the strongest shallow control remains below the DESG scorer. A raw-text rubric LLM judge reaches 0.3451, and a dev-selected text-embedding kNN risk baseline reaches 0.6006. DESG-StateRisk, a ridge harmful-risk head over DESG-extracted clinical-state transitions with alpha and threshold selected only on the development split, reaches 0.6494 macro-F1, the strongest point estimate in this audit. A paired bootstrap comparison with TextEmbedding-Risk gives a 0.0488 macro-F1 difference with a 95% interval of $[-0.0316, 0.1285]$, so we treat this as primary artifact-controlled evidence for useful state-transition signal, not as a statistically conclusive superiority claim.

The DESG ablations show that the gain is not coming from a single superficial state field. Post-state-only scoring reaches 0.5986, delta-emotion-only reaches 0.5085, delta-cognition-only reaches 0.5261, and removing asymmetric distance reaches 0.6451. The full DESG-StateRisk row is the best result, while the small gap to the no-asymmetry ablation suggests a measured interpretation: most of the clean-benchmark signal comes from the structured clinical-state transition representation, with asymmetric distance adding a modest additional improvement. This is the paper's strongest artifact-controlled positive result.

4.5 Selected-Direction Stress Audit

We also retain a label-origin-balanced stress audit as a boundary diagnostic. This stress set removes provenance and lexical shortcuts by construction and uses an ordinal selected-direction cue. The artifact controls collapse on this stress set: metadata-only and shallow bag-of-words both fall to 0.1667 macro-F1. The direct DeepSeek judge retains a strong raw-text signal, reaching 0.7387 held-out macro-F1 on the API-backed test split, while current cached-state template scorers are not designed for this selected-direction input format: ConcatANN and DESG-CORE reach 0.1667 and 0.2698. A stress-calibrated state ridge improves to 0.4558 but remains below the direct judge. We treat this result as a stress-specific input-structure diagnostic rather than the main benchmark claim: it shows that selected-direction parsing and single-response risk detection are different regimes, while the clean single-response matched benchmark shows that DESG state-transition risk scoring remains strongest when each sample contains one matched response.

4.6 Clinical-State and Mechanism Diagnostics

Clinical-state ablations provide diagnostic evidence about which feature groups are predictive for ConcatANN performance. Semantic-only scoring drops to 0.6559 macro-F1, while clinical-only scoring reaches 0.8755. This suggests that semantic content alone is weaker than affective and cognitive-distortion state features for this benchmark. The full ConcatANN model still performs better

Table 4. Label-origin-balanced selected-direction stress audit. The stress set removes easy metadata and lexical shortcuts and tests a different input structure from the clean single-response matched benchmark. Direct raw-text judging retains signal on this stress format, while current cached-state template scorers are not optimized for the selected-direction cue.

Source or method	N/coverage	Macro-F1	Accuracy	Interpretation
Metadata-only control	–	0.1667	–	Provenance shortcut removed
Shallow BoW control	–	0.1667	–	Lexical shortcut removed
Direct judge	36/36	0.7387	0.7500	Raw-text judging retains stress signal
State ConcatANN	180/180	0.1667	–	Boundary for current cached-state template path
State DESG-CORE	180/180	0.2698	–	Selected-direction parsing remains open
Stress-calibrated State Ridge	36	0.4558	0.4722	Train/dev calibration recovers partial signal but stays below direct judging

than clinical-only scoring, indicating that semantic context remains useful when combined with clinical-state features.

The largest individual drops come from removing Mental Filter, valence, and arousal. Removing Mental Filter reduces macro-F1 to 0.8004, corresponding to a +0.1199 drop from the full ConcatANN score. Removing valence and arousal reduces macro-F1 to 0.8168 and 0.8191, respectively. These diagnostic ablations are consistent with the DESG design assumption that text semantics alone is not enough for the constructed task.

The experimental evidence supports three bounded conclusions. First, general evaluator baselines and the direct LLM judge baseline are protocol-dependent: their relative performance changes across the constructed protocol, the selected-direction stress cue, and the clean single-response risk audit. Second, the clinical-state representation is highly informative on the constructed split, and the clean matched benchmark gives DESG state-transition risk scoring the strongest point estimate against metadata, lexical, surface, rubric-LLM, and text-embedding risk baselines under context-level splits. Third, DESG is best read as a structured diagnostic audit framework: it provides useful clinical-state evidence and inspectable audit traces, while the reported controls make leakage, shortcut pressure, and stress-set boundaries explicit rather than hidden.

5 Conclusion

Clinical dialogue evaluation needs to look beyond surface empathy, black-box judging, and unexamined benchmark leaderboards. We introduced DESG as a structured offline audit framework that converts dialogue windows into clinical states, optionally represents them as directed state graphs, and scores clinical direction with asymmetric state-transition features. On the original constructed 3×1000 diagnostic benchmark, DESG-Ensemble is the top-performing classifier by point estimate, though its small margin over ConcatANN should not be treated as a standalone significance claim. On the clean single-response matched benchmark, DESG-StateRisk is also the strongest harmful-risk detector by point estimate, ahead of metadata, lexical, surface, rubric-LLM, and text-embedding risk baselines under context-level splits.

The broader finding is methodological and diagnostic. Under the original constructed protocol, direct and released evaluator baselines often under-detect harmful directionality wrapped in supportive language, while cached clinical-state methods are highly predictive. The artifact audits show why ordinary leaderboards are not enough: metadata and lexical probes reveal construction shortcuts that would otherwise remain invisible. The clean matched benchmark turns that audit lesson into a stronger protocol by reporting source-label balance, context-level splits, duplicate checks, response-length matching, shared anchors, shallow controls, raw-text baselines, and DESG ablations together. DESG is therefore a practical audit framework rather than just a scoring model.

The benchmark remains an offline evaluator benchmark, not a clinical trial, and DESG should be used for auditing, dataset construction, red-team triage, and model comparison rather than autonomous diagnosis or treatment decisions. The selected-direction stress audit also shows that input structure matters: direct judging can use an ordinal selected-direction cue that current cached-state template scorers were not designed to parse. This boundary clarifies where future state extractors and scoring heads should improve. Implicit-sycophancy evaluation is therefore better treated as an auditable clinical-direction problem than as a generic response-quality judgment. DESG and the clean matched benchmark together provide a reusable protocol for testing whether an evaluator detects harmful direction beyond surface empathy while controlling for metadata, lexical, and structural shortcuts.

6 Ethics Statement

This work is limited to offline evaluation and red-team auditing of psychological dialogue systems. DESG is not a diagnostic, therapeutic, triage, or crisis-response system, and its outputs must not replace clinicians, counselors, crisis workers, or other qualified human decision makers. Any deployment-facing use would need prospective human review, domain-specific safety protocols, and institutional oversight.

The supported scope is limited to research auditing over saved dialogue windows, dataset debugging, evaluator comparison, and diagnostic pattern analysis.

Clinical diagnosis, treatment recommendation, and autonomous triage remain unsupported.

The benchmarks are built from existing research dialogue sources and generated counterfactual windows. They contain mental-health and crisis-oriented content, so examples need content warnings and should not be presented as advice. The original constructed benchmark is framed as a diagnostic protocol with reported artifact controls. The clean single-response matched benchmark adds stronger leakage controls, including context-level splits, source-label balance, duplicate checks, response-length matching, and shallow shortcut probes. It remains an offline evaluation resource rather than a clinical-label dataset for deployment. Public releases should follow upstream dataset licenses and avoid exposing administrative metadata or construction fields that enable trivial label recovery.

Large language models are used only as structured state extractors and direct LLM judge baselines in cached offline experiments; they do not interact with users, make clinical decisions, or trigger interventions. The method makes evaluator behavior more inspectable, but does not establish clinical efficacy or real-world mental-health deployment safety. Future annotation or deployment-facing validation should follow institutional ethics review.

Acknowledgments. We also thank Shenzhen University Stability Support Project (grant no. 20231128113233002).

Disclosure of Interests. The authors have no competing interests to declare that are relevant to the content of this article. The benchmark and metrics in this paper are intended for offline evaluation and red-team analysis, not for autonomous clinical diagnosis or treatment.

References

1. Beck, A.T., Rush, A.J., Shaw, B.F., Emery, G.: Cognitive Therapy of Depression. Guilford Press (1979)
2. Braun, J.D., Strunk, D.R., Sasso, K.E., Cooper, A.A.: Therapist use of socratic questioning predicts session-to-session symptom change in cognitive therapy for depression. Behav. Res. Ther. **70**, 32–37 (2015). https://doi.org/10.1016/j.brat.2015.05.004
3. Chen, G.H., Chen, S., Liu, Z., Jiang, F., Wang, B.: Humans or LLMs as the judge? A study on judgement bias. In: Proceedings of the 2024 Conference on Empirical Methods in Natural Language Processing, pp. 8301–8327. Association for Computational Linguistics (2024). https://doi.org/10.18653/v1/2024.emnlp-main.474
4. Chen, Y., Yan, S., Liu, S., Li, Y., Xiao, Y.: EmotionQueen: a benchmark for evaluating empathy of large language models. In: Findings of the Association for Computational Linguistics: ACL 2024, pp. 2149–2176. Association for Computational Linguistics (2024). https://doi.org/10.18653/v1/2024.findings-acl.128
5. Chiang, C.H., Lee, H.y., Lukasik, M.: TRACT: regression-aware fine-tuning meets chain-of-thought reasoning for LLM-as-a-judge. In: Proceedings of the 63rd Annual Meeting of the Association for Computational Linguistics (Volume 1: Long Papers),

pp. 2934–2952. Association for Computational Linguistics (2025). https://doi.org/10.18653/v1/2025.acl-long.147
6. D'Souza, J., Babaei Giglou, H., Münch, Q.: YESciEval: robust LLM-as-a-judge for scientific question answering. In: Proceedings of the 63rd Annual Meeting of the Association for Computational Linguistics (Volume 1: Long Papers), pp. 13749–13783. Association for Computational Linguistics (2025). https://doi.org/10.18653/v1/2025.acl-long.675
7. Elliott, R., Bohart, A.C., Watson, J.C., Greenberg, L.S.: Empathy. Psychotherapy **48**(1), 43–49 (2011). https://doi.org/10.1037/a0022187
8. Gould, M.S., Kalafat, J., Harris Munfakh, J.L., Kleinman, M.: An evaluation of crisis hotline outcomes part 2: suicidal callers. Suicide Life Threat. Behav. **37**(3), 338–352 (2007). https://doi.org/10.1521/suli.2007.37.3.338
9. Gu, J., et al.: A survey on LLM-as-a-judge. Innovation **7**(6), 101253 (2026). https://doi.org/10.1016/j.xinn.2025.101253
10. Hofmann, S.G., Asnaani, A., Vonk, I.J.J., Sawyer, A.T., Fang, A.: The efficacy of cognitive behavioral therapy: a review of meta-analyses. Cogn. Ther. Res. **36**(5), 427–440 (2012). https://doi.org/10.1007/s10608-012-9476-1
11. Inkster, B., Sarda, S., Subramanian, V.: An empathy-driven, conversational artificial intelligence agent (Wysa) for digital mental well-being: real-world data evaluation mixed-methods study. JMIR Mhealth Uhealth **6**(11), e12106 (2018). https://doi.org/10.2196/12106
12. Kim, S., et al.: Prometheus 2: an open source language model specialized in evaluating other language models. In: Proceedings of the 2024 Conference on Empirical Methods in Natural Language Processing, pp. 4334–4353. Association for Computational Linguistics (2024). https://doi.org/10.18653/v1/2024.emnlp-main.248
13. Lee, D., Hwang, Y., Kim, Y., Park, J., Jung, K.: Are LLM-judges robust to expressions of uncertainty? investigating the effect of epistemic markers on LLM-based evaluation. In: Proceedings of the 2025 Conference of the Nations of the Americas Chapter of the Association for Computational Linguistics: Human Language Technologies (Volume 1: Long Papers), pp. 8962–8984. Association for Computational Linguistics (2025). https://doi.org/10.18653/v1/2025.naacl-long.452
14. Leng, Y., et al.: Praetor: a fine-grained generative LLM evaluator with instance-level customizable evaluation criteria. In: Proceedings of the 63rd Annual Meeting of the Association for Computational Linguistics (Volume 1: Long Papers), pp. 10386–10418. Association for Computational Linguistics (2025). https://doi.org/10.18653/v1/2025.acl-long.513
15. Li, A., Lu, Y., Song, N., Zhang, S., Ma, L., Lan, Z.: Understanding the therapeutic relationship between counselors and clients in online text-based counseling using LLMs. In: Findings of the Association for Computational Linguistics: EMNLP 2024, pp. 1280–1303. Association for Computational Linguistics (2024). https://doi.org/10.18653/v1/2024.findings-emnlp.69
16. Li, D., et al.: From generation to judgment: opportunities and challenges of LLM-as-a-judge. In: Proceedings of the 2025 Conference on Empirical Methods in Natural Language Processing, pp. 2757–2791. Association for Computational Linguistics (2025). https://doi.org/10.18653/v1/2025.emnlp-main.138
17. Li, J., Sun, S., Yuan, W., Fan, R.Z., Zhao, H., Liu, P.: Generative judge for evaluating alignment. In: The Twelfth International Conference on Learning Representations (2024). https://openreview.net/forum?id=gtkFw6sZGS
18. Li, Y., Yao, J., Bunyi, J.B.S., Frank, A.C., Hwang, A.H.C., Liu, R.: CounselBench: a large-scale expert evaluation and adversarial benchmarking of large language

models in mental health question answering. arXiv preprint arXiv:2506.08584 (2025)
19. Li, Z., et al.: Split and merge: aligning position biases in LLM-based evaluators. In: Proceedings of the 2024 Conference on Empirical Methods in Natural Language Processing, pp. 11084–11108. Association for Computational Linguistics (2024). https://doi.org/10.18653/v1/2024.emnlp-main.621
20. Liu, S., et al.: Towards emotional support dialog systems. In: Proceedings of the 59th Annual Meeting of the Association for Computational Linguistics and the 11th International Joint Conference on Natural Language Processing (Volume 1: Long Papers), pp. 3469–3483. Association for Computational Linguistics (2021). https://doi.org/10.18653/v1/2021.acl-long.269
21. Na, H., et al.: A survey of large language models in psychotherapy: current landscape and future directions. In: Findings of the Association for Computational Linguistics: ACL 2025, pp. 7362–7376. Association for Computational Linguistics (2025). https://doi.org/10.18653/v1/2025.findings-acl.385
22. Nguyen, V.C., et al.: Do large language models align with core mental health counseling competencies? In: Findings of the Association for Computational Linguistics: NAACL 2025, pp. 7503–7526. Association for Computational Linguistics (2025). https://doi.org/10.18653/v1/2025.findings-naacl.418
23. Panickssery, A., Bowman, S.R., Feng, S.: LLM evaluators recognize and favor their own generations. In: Advances in Neural Information Processing Systems (2024). https://arxiv.org/abs/2404.13076
24. Park, J., Jwa, S., Meiying, R., Kim, D., Choi, S.: OffsetBias: leveraging debiased data for tuning evaluators. In: Findings of the Association for Computational Linguistics: EMNLP 2024, pp. 1043–1067. Association for Computational Linguistics (2024). https://doi.org/10.18653/v1/2024.findings-emnlp.57
25. Rashkin, H., Smith, E.M., Li, M., Boureau, Y.L.: Towards empathetic open-domain conversation models: a new benchmark and dataset. In: Proceedings of the 57th Annual Meeting of the Association for Computational Linguistics, pp. 5370–5381. Association for Computational Linguistics (2019). https://doi.org/10.18653/v1/P19-1534
26. Reimers, N., Gurevych, I.: Sentence-BERT: sentence embeddings using Siamese BERT-networks. In: Proceedings of the 2019 Conference on Empirical Methods in Natural Language Processing and the 9th International Joint Conference on Natural Language Processing (EMNLP-IJCNLP), pp. 3982–3992. Association for Computational Linguistics (2019). https://doi.org/10.18653/v1/D19-1410
27. Rnic, K., Dozois, D.J.A., Martin, R.A.: Cognitive distortions, humor styles, and depression. Eur. J. Psychol. **12**(3), 348–362 (2016). https://doi.org/10.5964/ejop.v12i3.1118
28. Rogers, C.R.: The necessary and sufficient conditions of therapeutic personality change. J. Consult. Psychol. **21**(2), 95–103 (1957). https://doi.org/10.1037/h0045357
29. Rose, A.J.: Co-rumination in the friendships of girls and boys. Child Dev. **73**(6), 1830–1843 (2002). https://doi.org/10.1111/1467-8624.00509
30. Russell, J.A.: A circumplex model of affect. J. Pers. Soc. Psychol. **39**(6), 1161–1178 (1980). https://doi.org/10.1037/h0077714
31. Shan, G., et al.: Dental diagnosis from x-ray panoramic radiography images: a dataset and a hybrid framework. In: Chinese Conference on Pattern Recognition and Computer Vision, pp. 234–248 (2024). https://doi.org/10.1007/978-981-97-8496-7_17

32. Shi, L., Ma, C., Liang, W., Diao, X., Ma, W., Vosoughi, S.: Judging the judges: a systematic study of position bias in LLM-as-a-judge. In: Proceedings of the 14th International Joint Conference on Natural Language Processing and the 4th Conference of the Asia-Pacific Chapter of the Association for Computational Linguistics, pp. 292–314. The Asian Federation of Natural Language Processing and The Association for Computational Linguistics (2025). https://doi.org/10.18653/v1/2025.ijcnlp-long.18
33. SungJoo: CRADLE-Dialogue: crisis-response dialogue dataset. Hugging Face dataset (2026). https://huggingface.co/datasets/SungJoo/Cradle-Dialogue, dataset card
34. Vaidyam, A.N., Wisniewski, H., Halamka, J.D., Kashavan, M.S., Torous, J.B.: Chatbots and conversational agents in mental health: a review of the psychiatric landscape. Can. J. Psychiatry **64**(7), 456–464 (2019). https://doi.org/10.1177/0706743719828977
35. Wang, X., Zhang, J., Zhang, G., Guo, H.: Feel the difference? A comparative analysis of emotional arcs in real and LLM-generated CBT sessions. In: Findings of the Association for Computational Linguistics: EMNLP 2025, pp. 19999–20017. Association for Computational Linguistics (2025). https://doi.org/10.18653/v1/2025.findings-emnlp.1089
36. Wataoka, K., Takahashi, T., Ri, R.: Self-preference bias in LLM-as-a-judge. arXiv preprint arXiv:2410.21819 (2024)
37. Watts, I., Gumma, V., Yadavalli, A., Seshadri, V., Swaminathan, M., Sitaram, S.: PARIKSHA: a large-scale investigation of human-LLM evaluator agreement on multilingual and multi-cultural data. In: Proceedings of the 2024 Conference on Empirical Methods in Natural Language Processing, pp. 7900–7932. Association for Computational Linguistics (2024). https://doi.org/10.18653/v1/2024.emnlp-main.451
38. Xie, H., Chen, Y., Xing, X., Lin, J., Xu, X.: PsyDT: using LLMs to construct the digital twin of psychological counselor with personalized counseling style for psychological counseling. In: Proceedings of the 63rd Annual Meeting of the Association for Computational Linguistics (Volume 1: Long Papers), pp. 1081–1115. Association for Computational Linguistics (2025). https://doi.org/10.18653/v1/2025.acl-long.55
39. Zhai, W., et al.: MentalGLM series: explainable large language models for mental health analysis on Chinese social media. In: Proceedings of the 2025 Conference on Empirical Methods in Natural Language Processing, pp. 13588–13603. Association for Computational Linguistics (2025). https://doi.org/10.18653/v1/2025.emnlp-main.686
40. Zhang, C., et al.: CPsyCoun: a report-based multi-turn dialogue reconstruction and evaluation framework for Chinese psychological counseling. In: Findings of the Association for Computational Linguistics: ACL 2024, pp. 13947–13966. Association for Computational Linguistics (2024). https://doi.org/10.18653/v1/2024.findings-acl.830
41. Zhang, M., et al.: CBT-bench: evaluating large language models on assisting cognitive behavior therapy. In: Proceedings of the 2025 Conference of the Nations of the Americas Chapter of the Association for Computational Linguistics: Human Language Technologies (Volume 1: Long Papers), pp. 3864–3900. Association for Computational Linguistics (2025). https://doi.org/10.18653/v1/2025.naacl-long.196

42. Zhang, Q., et al.: Crowd comparative reasoning: Unlocking comprehensive evaluations for LLM-as-a-judge. In: Proceedings of the 63rd Annual Meeting of the Association for Computational Linguistics (Volume 1: Long Papers), pp. 5059–5074. Association for Computational Linguistics (2025). https://doi.org/10.18653/v1/2025.acl-long.252
43. Zhang, T., Kishore, V., Wu, F., Weinberger, K.Q., Artzi, Y.: BERTScore: evaluating text generation with BERT. In: International Conference on Learning Representations (2020). https://openreview.net/forum?id=SkeHuCVFDr
44. Zhao, H., et al.: ESC-eval: evaluating emotion support conversations in large language models. In: Proceedings of the 2024 Conference on Empirical Methods in Natural Language Processing, pp. 15785–15810. Association for Computational Linguistics (2024). https://doi.org/10.18653/v1/2024.emnlp-main.883

Author Index

H. Chen et al. (Eds.): IJCAI 2026, CCIS 3120, pp. 225–226, 2027.
https://doi.org/10.1007/978-981-92-4529-1

MIX
Papier aus verantwortungsvollen Quellen
Paper from responsible sources
FSC® C105338

If you have any concerns about our products,
you can contact us on
ProductSafety@springernature.com

In case Publisher is established outside the EU,
the EU authorized representative is:
Springer Nature Customer Service Center GmbH
Europaplatz 3, 69115 Heidelberg, Germany

Printed by Libri Plureos GmbH
in Hamburg, Germany